EXTRAORDINARY ENGINEERS

Volume 1

DR. J. A. SANCHEZ
EXTRAORDINARY ENGINEERS

Extraordinary Engineers
By Dr. J. A. Sanchez

Copyright© 2023 Dr. J. A. Sanchez

Extraordinary Engineers:
Extraordinary Engineers Volume 1
Female Engineers of This Day and Time

Radiant Pearls Publishing
2023

All journeys in this book are contributions of the extraordinary engineers.

Printed in the U.S.A.

ISBN:979-8-9877073-7-1(paperback)
ISBN:979-8-9877073-8-8(hardcover)
ISBN:979-8-9877073-6-4 (ebook)

All rights reserved.

This book is protected under the copyright laws of the United States of America. Any reproduction or unauthorized use of the material or artwork contained herein is prohibited without the express written permission of Dr. J. A. Sanchez.

No part of this publication may be reproduced or transmitted in any form or by any means, electronic or mechanical, without permission in writing from the author.

Request for permission to make copies of any part of this work should be directed to the following email: theextraordinaryengineers@gmail.com.

Cover art done by Get Covers https://getcovers.com/

Editing done by Jeanette Crystal Bradley and Tiffany Vakilian

Scriptures taken from the New King James Version. Copyright© 1982 by Thomas Nelson, Inc. Used by permission. All rights reserved.

Scripture quotations from The Authorized (King James) Version. Rights in the Authorized Version in the United Kingdom are vested in the Crown. Reproduced by permission of the Crown's patentee, Cambridge University Press.

CONTENTS

Acknowledgements................................v
Introduction....................................vii
Foreword..ix
Extraordinary Engineers..........................1

See Syd Soar
Coauthored by Sydney Hamilton...................3

How a High School Class Gave Me the Courage to Major in Mechanical Engineering
Coauthored by Libby Brooks.....................17

From Chemical Engineer to Chief Wellness Engineer—A Path Full of Surprises
Coauthored by Lennis Perez.....................23

STEM with Jen
Coauthored by Jen Patel........................33

The People Engineer®
Coauthored by Brianne C. Martin................41

Getting From One Place to Another
Coauthored by Vanessa Eslava...................65

A PlastChick
Coauthored by Lynzie Nebel.....................69

Whatever You Do, Don't Stop, and Don't Give Up!
Coauthored by Michelle Vargas 77

A Journey of Discovery: Finding Purpose in Inspiring Others
Coauthored by Kelly Kloster Hon 93

Extraordinary Engineers
Coauthored by Dr. J. A. Sanchez 105

Questions and Answers with the Extraordinary Engineers . . .117

Cuentame con Sonia
Contributed by Sonia Camacho 119

Aspiring Engineer
Contributed by Madalyn Nguyen 127

AdAstraSu
Contributed by Susan Martinez 137

Reinvented
Contributed by Caeley Looney 145

NoireSTEMinist®
Contributed by Dr. Carlotta Berry 153

Watch Me STEM
Contributed by Paula Garcia Todd 159

All Things Chemical Engineering
Contributed by Alexis Enacopol. 167

The Space Latina
Contributed by Zaida Hernandez 177

I Was Born To Be An Engineer
Contributed by Mandy Zhao 183

A Life In Problem Solving
Contributed by Dr. Susan Rogers 187

ACKNOWLEDGEMENTS

First and foremost, I want to thank God for allowing me to be a part of bringing this book to you. He has been so kind and gracious with me every step of the way. It is an honor to partner with Him on this project.

My husband, Mike, thank you for always supporting me and loving me. I love you forever.

To my daughter, Lillyan, you have taught me so much and helped me grow in ways I never imagined. You are the best daughter in the whole world, and I am blessed to be your mommy! You constantly blow me away with your kindness, generosity, and love. You are so thoughtful, smart, and amazing! You are a leader for this generation. I love you so much, always and forever, baby girl! God has a great plan and such a bright future for you! Keep seeking Him.

To my mom, Debbie, thank you for your constant love and support. Thank you for being my biggest cheerleader. When I look back at my life, you have been the one constant person cheering me on and loving me through it all. I am so thankful God chose you to be my mom and Lilly's grammy. Your love and support have been a huge part of bringing this book to life, and I am so grateful for you.

To my bestie, Jeanette, the one who knows how to have my kind of fun! Thank you for all the love, laughs and delicious dessert dates. Your love, encouragement, and support have been priceless. I appreciate you so much. You are the best bestie, partner, and sister ever!

To all of the Extraordinary Engineers featured in this book, thank you for giving of your time, talent, and treasure, and for sharing your story. You all do so much to inspire and support our next generation of engineers, and that is invaluable!

To my mentors, Stefan Rentsch and Jobina Gonsalves, thank you for pouring into me. Your time, guidance, experiences, and knowledge have had such a huge impact on me, and you have helped me realize that I can be and do so much more than I ever thought I could.

Thank you, Jesus, for your unending, pure love for me and all of mankind. You are what it is all about.

Life Verse: Seek first the Kingdom of God and His righteousness and all these things shall be added unto you. Matthew 6:33 (NKJV)

INTRODUCTION

Over the past eighteen years of being in the tech industry, I have had the opportunity to collaborate with many organizations, engineers, and professionals in STEM (Science, Technology, Engineering, and Math), and I walk away so inspired! I would hear the stories of the road less traveled by women in careers I hadn't thought much about, which was so eye-opening. I remember being part of a STEM summit and hearing a woman, who was an engineer that worked for the Corvette car company, talk about the seats in a Corvette and how every time she got in and out, anyone could see right up her dress. Clearly, the design team hadn't considered women's attire when designing the seats. She had the opportunity to be on the design team and could bring her voice to the table which made a huge impact on how the seats are made on every model. I was blown away by hearing how something I never considered has made a huge impact on society. This is just one of the billions of reasons why it is so important to have women in engineering. We experience life in different ways than men do. Women bring a unique perspective; we think differently from our male colleagues. Often, the approach to solving the same problem comes from a different perspective; women tend to find a solution using a different route, not that one is right and one is wrong, but what we bring is different. Absence of women in this sector means we (engineering as a whole) are losing out on many valuable ideas, solutions, and opportunities. Did you know that according to the U.S.

Department of Labor Women's Bureau, women make up only 15 percent of the engineering workforce? Something seems off balance, right? That means this world is being robbed of the very solutions that could help make it a better place to live.

This book will provide insight into the lives of Female Engineers of this day and time and their experiences, I know firsthand that having the visibility of those who are further ahead in the areas of life that we want to grow in, is incredibly important. Those who did the hard thing and went against the grain. Those who had the courage to pursue their dreams against all odds. Those who will show you the path they walked, the risks they took, and give guidance from their perspective. It could be the very thing that you need to change the trajectory of your life. I hope to bring you to a place that opens your eyes to the world of opportunities available to you. The opportunities already there and the ones waiting to be created, maybe even by you. I hope you learn some valuable lessons from those who have taken the steps you want to take and feel inspired to be the change you want to see.

What you bring matters; your ideas, your thoughts, and your plans—engineering needs you!

FOREWORD

For girls in STEM, exposure to role models involved in STEM has the power to pique their interest and increase the likelihood that they will pursue STEM interests. Girls who see themselves in others imitate the traits with which they most identify.

Representation is powerful, and it matters. When a child sees an experience that represents their own which focuses on diversity or includes black, indigenous, and people of color characters, it has the power to impact how they feel about themselves and others. Representation ensures diverse viewpoints and perspectives can be found or depicted in society.

Justina and I met at a Society of Women Engineers (SWE), San Diego Professional Section open house, where we made an immediate connection since Justina was a product safety engineer, and my daughter Madalyn's robotics team was known as a safety team. Since that day, Justina has been Madalyn's most prominent advocate, aside from her family. Justina supported her by creating an internship position for Madalyn when she was finishing her freshman year. Thereafter, she made every effort to be present at all of Madalyn's speaking engagements, from WE19 to WE Local San Diego, as the keynote speaker. When Madalyn received the 10 News Leadership Award, it was Justina who was there supporting her and being interviewed by the news on Madalyn's behalf. Every student needs a mentor and role model like Justina.

Growing up, Justina Sanchez never had a STEM role model. She did not realize engineering was a field for women. When she first started at TÜV SÜD, she was just looking for a decent job with benefits. After working there for two years, she began to realize there was a whole world of engineering that she never knew existed. She saw it as fun and exciting, since to her, no two days were alike. Once she had a glimpse of what her life would be like if she were an engineer, her eyes and heart would light up. The more she thought about it, the more her heart would leap out of her chest with excitement. When she inquired about becoming an engineer, her boss gave her the nudge to go to school to be certified in electronic engineering. With hard work and sacrifices, she was promoted to Product Safety Test Technician, and eventually, after gaining on-the-job experience, she was promoted to Product Safety Engineer.

After eighteen years at TÜV SÜD, Justina has climbed the promotional ladder and is a successful engineer, living her best life. With her vast experience in an accredited lab, she was given the opportunity to be an auditor. Thus, her journey to becoming a quality engineer is unconventional, to say the least. Justina extraordinarily became an engineer despite lacking role models and being nearly surrounded by males in her classes and educational institute. This experience led her to become an extraordinary role model for countless youth interested in engineering.

Since realizing her success, she has been on a mission to be the STEM role model that she never had. Compiling this book is Justina's passion project to promote representation in engineering. All girls should know they can aspire to become role models and that all paths are possible.

It has been a privilege to be a part of Justina's engineering journey since meeting her in 2018 and watching her evolve into an amazing engineer, STEM advocate, and compassionate human being. Justina is one of the most active and committed members of the SWE San Diego Outreach Committee. We are constantly volunteering together at outreach events as we both truly believe in SWE's vision of a world with gender parity in engineering and technology. We are both catalysts for change for women in engineering and technology. It is my great honor to introduce <u>Extraordinary Engineers – Female Engineers of this Day and Time</u>, a showcase of amazing, diverse engineers as role models. It is my privilege to introduce Justina Sanchez as the ultimate STEM role model for the next generation of engineers and innovators.

—Tracy Nguyen, O.D.
SWE San Diego Outreach Director

EXTRAORDINARY ENGINEERS

SEE SYD SOAR

Coauthored by Sydney Hamilton:
Rocket Scientist and Engineering Manager

The Background

"I am lucky that whatever fear I have inside me, my desire to win is always stronger."

—*Serena Williams*,
Former No. 1 World Ranked Tennis Player.

I have conquered rejection, vanquished imposter syndrome, and survived failure. Hi, my name is Sydney Hamilton, and I am a superhero. My superpowers are all the attributes that make me unique. I am here not only to change the world but the galaxy as well.

I have had an exciting career as an aerospace engineer and engineering manager. Outside of work, I have had diverse aerospace-related experiences ranging from training as a commercial astronaut to becoming an aquanaut by diving into an underwater habitat to live underwater for twenty-four hours! Through my community outreach, I have even been honored with a life-sized 3D-printed statue of myself that was displayed at the Smithsonian.

And this is where my story began:

I remember the moment I knew I would fly one day. I was six years old and living in Texas. I had spent hours jumping off the dining room chair, convinced I had finally gained enough "flight time" to take off. So I sprinted up the wooden steps of my home, searching frantically for my cape. Once I found it, I ran outside to my launch pad (our red-paved driveway) and mentally prepared for the takeoff. I stood with my feet together, a blanket cape tied around my neck, and my arms straight out. I was ready. I shut my eyes tight and focused on lifting off the ground, but I remained rooted no matter how hard I tried. I tried focusing on blasting off from my hands, then from my feet, jumping as high as possible, or doing anything to get off the ground. Feeling defeated, I feared I might never fly. After about thirty minutes of failed attempts, my mom came outside and suggested I try to fly another way. She suggested a jet pack, an airplane, or even a rocket ship.

"No, Mom. I want to fly!" I replied.

Not once did my parents tell me I couldn't fly. They reminded me, "Where there is a will, there is a way." I always came back to believing that there was a way that I could fly. So, it is no surprise I grew up interested in soaring the skies. As an aerospace engineer, I am flying, just slightly different than I initially imagined.

The Challenges

"I thrive on obstacles. If I'm told that it can't be done, then I push harder."

—Issa Rae,
Actress, Writer, Producer, and Comedian.

Growing up, my dad would always call me his little engineer. He would highlight the seemingly insignificant accomplishment as if it were worthy of the Nobel Prize. I've always been interested in math, building, and being creative. While spell-check must work extra hard when I write, math has always made sense to me. My local library had science experiments you could check out and do at home. I loved doing science experiments at home. I even started volunteering at the library to be the first to check them out.

Near the end of my senior year of high school, I took an aptitude test to help me determine my future career path. I was eager to see what opportunities matched my skillsets. I was thrilled when the results indicated that I should pursue a career as an engineer. Excited by this prospect, I proudly shared the news with one of my teachers when he asked about my results. However, I was surprised when my English teacher responded with his dismissive comment, "Isn't that too hard for someone like you?" This comment hit me hard, and I began to doubt my abilities. I felt defeated and unsure of myself. But a few days later, my orchestra teacher pulled me aside and asked what was wrong. I told her the story, and she responded with a smirk and a question, "Since when do you listen to anyone?" I looked at her and smiled; she was right. Her words reminded me of my determination and resilience, and I realized that I had always forged my path in life. "Where there is a will, there is a way." I embraced this same spirit and pursued my dream of becoming an engineer. I am grateful to my orchestra teacher.

Despite anyone's doubts, I went to college to become an engineer. Through the Atlanta University Consortium Dual Degree program, I went to Spelman College and the University of Michigan, where I majored in mathematics and aerospace engineering, respectively.

Spelman College is an all-women's historically Black college or university (HBCU) in Atlanta, Georgia, and was the first school in the United States where Black women could gain a higher education. I spent three life-changing years learning how to be a confident leader, love myself exactly as I am, and gain an excellent education supported by teachers who looked like me for the first time. They taught me to challenge what others say and to be open to being challenged about how I think. It is okay to change your mind with new information. My professors and Spelman sisters inspire me to this day with their vigor for pursuing growth and their passion.

After Spelman, I went to the University of Michigan in Ann Arbor, MI, where I studied aerospace engineering. It was exciting to attend one of the country's top engineering schools. This is where I learned the power of teamwork. However, it was also very overwhelming, and I had a tough time fitting in initially. The school of engineering was an entirely new challenge, and I did not have a support group like my Spelman sisters when I started. Eventually, I got involved in student organizations, joined a lab, and made friends in class and with my roommates. What seemed to be an overwhelming new experience soon became a place where I tackled the most challenging problems and persevered. Excitingly, immediately after graduation, I started with a full-time offer at a great company.

> *"Whatever your reason for holding on to resentments, I know this for sure: There is none worth the price you pay in lost time."*
>
> —***Oprah Winfrey***, Talk Show host, Producer, Actress, Author, and Philanthropist.

The University of Michigan honored my accomplishments in the industry at their Aerospace Department's Centennial event a couple of years after I graduated. I was indescribably proud and excited to share my experiences and achievements. When back in Texas visiting my family, I decided to visit my high school. I was thrilled that the English teacher who told me engineering was "too hard" was still there. I proudly marched into his classroom, looked him in the eyes, and announced, "My name is Sydney Hamilton, and I am a successful aerospace engineer." You could fill the room with my ego at this moment. He looked back at me and paused. Then he said, "I am so sorry, and you are?" I was in utter disbelief. Feeling less confident, I said, "I am Sydney Hamilton, one of your former students." Humility and embarrassment hit me like a ton of bricks as I explained who I was and where I used to sit. All these years, I held on to the hurt from a man who did not even remember I existed.

As I left his room, I realized that this person probably had no idea how much his words had affected me. It was a wake-up call to realize that I had wasted time and energy on someone who did not care enough to remember me. This experience taught me not to take criticism too seriously when it comes from someone I would not have sought advice from in the first place.

The Why

"I've learned that making a 'living' is not the same as 'making a life.'"

—**Maya Angelou**, Memoirist,
Popular Poet, and Civil Rights Activist.

When I was younger, I struggled to understand that I, too, could be a princess because none of them looked like me. No matter how often my dad called me his "little princess" or I heard him call my mom "his Queen," it did not feel like the rest of the world saw me in that same light. So, I decided to be someone to whom girls and young women could relate. I chose to be an encouraging voice in the lives of others. Every person has the right to know they can do or be anything they want. I want to inspire everyone to rid themselves of negative voices because they have seen someone like them reach their goals too.

The Career / Real World

"You never have to ask anyone's permission to lead. Just Lead."

– **Kamala Harris**, United States Vice President.

As an aerospace engineer and engineering manager at a Fortune 100 aerospace company in California, I have overseen the communication and coordination between customers, suppliers, and engineers on various space and commercial aircraft programs. I have the privilege of leading a talented team of engineers and am actively involved in building a global team by hiring and leading the next generation of global aerospace leaders. Seeing my team grow and thrive is extremely rewarding, and I am constantly motivated by the opportunity to positively impact each individual, as well as our industry.

When we think about engineering, we automatically think of hands-on work. However, for me, I was often at my desk. When I started as a design engineer, my computer became my canvas, where I created and refined designs for airplanes. We would have an engineering problem that I was responsible for solving. When I became an engineer, I hoped to avoid anything resembling my writing class. However, I quickly learned how essential writing and communication were to be successful as an engineer. I worked closely with my team, had many brainstorming sessions, and asked many questions. I was responsible for making our brainstorming session ideas come to life. I loved being an engineer because the work I did touched lives. Now, I lead projects that could change the aerospace industry. The best part is that I am bringing along the best and brightest to be a part of this remarkable masterpiece.

The Highlights

"Never be limited by other people's limited imagination."

—***Dr. Mae Jemison***, *Engineer, Physician, and Former NASA Astronaut.*

I have had a remarkably fruitful career journey, from working on the first commercial aircraft with mechanically folding wing tips, to jet-setting to different locations to repair aircraft to becoming an aerospace engineering manager at one of the largest aerospace engineering companies in the world. I have been able to start new employee groups and make positive impacts on the work I have done.

I have always been passionate about giving back and improving the lives of others. This desire to make a positive impact has been a part of my life since I was a child, and I am constantly inspired to find new ways to make a difference in my community and beyond. In 2021, I was honored to receive the Promise Award from the Space and Satellite Professionals International (SSPI)—an organization established in 1983—for my work in the aerospace industry. As the first African American to receive this award, I feel proud and grateful for the recognition. My work has also been featured on CBS's *Mission Unstoppable* television show, in *Marie Claire* magazine, and at the Smithsonian with a life-sized 3D-printed statue. I am humbled by the recognition and excited to continue contributing to the aerospace industry and the lives of many.

If someone had told me I would one day be going through a commercial astronaut training experience, I would have laughed in disbelief. I do not have a master's degree or military experience—things I thought were necessary to be selected to go to space. I had even convinced myself that it was not a realistic dream. However, through a traditional application process, I was contracted by Uplift Aerospace to join their Space+5 Astronaut Class. We are the first Web3 Astronaut Class. This opportunity has opened a whole new world for me and enabled me to inspire more young girls to pursue STEM careers and increase exposure to space opportunities.

Dream without bounds because you never know when "impossible" will become "I'm possible."

The Imposter

> *"Don't try to lessen yourself for the world; let the world catch up to you."*
>
> —***Beyonce Knowles-Carter***,
> *Global Award-Winning Artist and Actress.*

In college, I was told there would be no future job opportunities in the aerospace industry, which scared me. It made me think I would have done all this work in school only to struggle to find job opportunities after graduating from college. Upon looking at open job requisitions, reaching out to people in the industry, and talking to my university's career counselor, I learned that there are more opportunities than I ever imagined as an aerospace engineer. You can work on rockets, airplanes, submarines, and even cars. With this degree, you can make a variety of career choices.

Some people may try to convince you that your dreams are not enough, that they are too much, impossible to achieve, not suitable for you, or that you don't have what it takes to succeed. Remember, there are enough people in the world to tell you that you cannot do something or will not be able to accomplish your dreams. There is no need for you to be one of them. Believe in yourself even when no one else does—it is your future to determine, not theirs.

They say you are your own worst enemy, and it's true! Being a woman in Science, Technology, Engineering, and Math (STEM) is very rewarding but comes with challenges. It can be easy to doubt yourself. Imposter syndrome is a perceived fraud where people, often women and high achievers, doubt their capabilities despite their accomplishments and fear being discovered as frauds. Imposter syndrome can hinder your productivity, stall your motivation, and let your dreams stay asleep. I combat my imposter syndrome by having my Wing Team (a group of supportive people in my life) uplift me when I'm down or doubting myself. I also have mantras that help me through challenging times. One of my quotes is:

> *"See the world like an astronaut."*
>
> —**Sydney Hamilton**, *Rocket Scientist, Manager, STEM Advocate, CEO.*

This is a reminder to look at the big picture. When astronauts look back at Earth from space, they probably aren't thinking about the typo in that message or the toast they burned that morning, but are in awe of the change they are about to make. This helps me keep things in perspective. Shine bright because you are a star. The galaxy is vast, and there is enough room for every star to burn as brightly as they do.

The Advocacy

> *"Success isn't about how much money you make; it's about the difference you make in people's lives."*
>
> —**Michelle Obama**, *former First Lady, Lawyer, and Author.*

I am excited to expose more girls and young women to STEM (Science, Technology, Engineering, and Math) and STEAM (Science, Technology, Engineering, Art, and Math) career opportunities. Representation is critical to the future of STEM! If we build an inclusive team from the beginning, there will be less retrofitting or redoing to adjust to different abilities and lifestyles. I work to inspire the next generation of STEM leaders with my involvement as an American Association of the Advancement of Science (AAAS) If/Then Ambassador. Our motto is: "If she can see it, then she can be it." I created a See You Soar Scholarship that sent four girls to Space Camp for a week to expose them to more opportunities. My motivation is to be the representation I didn't have growing up; that way, all girls will know they can be princesses, engineers, CEOs, and so much more.

I am grateful for my opportunities to speak with students and share my story. I hope my story will inspire them to pursue their dreams. Even though I have given many speeches and presentations, including a TEDx talk at the London School of Economics, being a global keynote speaker, and hosting workshops for large corporations, I still have chilling moments before I get on stage. Yet, the feeling of making a positive impact and supporting others in their goals always overcomes the nerves. It is not about not being nervous; it is about being braver than you are.

In addition to my work as an engineering hiring manager, I actively support nonprofits and museums in various capacities, including serving as board president, volunteering, and speaking at their events. I also use my social media platforms to promote women in STEM. As a hiring manager, I work with my teams to ensure that we interview diverse candidates by attending conferences that showcase various candidates. A diverse team creates a more inclusive and positive work environment and helps us more effectively address the engineering challenges faced in the aerospace industry.

The Extracurricular

"You can't be hesitant about who you are."

—***Viola Davis***, *Oscar-Winning Actress.*

Life is all about balance. I adore spending time with my family. I have supportive parents, my sister is my best friend, and I am my niece and nephew's favorite auntie. I cannot give that up! But on the other hand, I am horizon-focused and excited to see where my career will take me. Trust me, you can have it all . . . just not all at once.

Having a balanced life will look differently for everyone. The balance will help you feel fulfilled and give you the space to show up in both your personal and professional lives. When you have a good balance, you can enjoy your time outside of work, pursue your hobbies and interests, and spend quality time with friends and family.

In my spare time, I love to scuba dive. It is fun to see the world from a different perspective—underwater. It is quiet and peaceful when I am diving. Fun Fact: Scuba diving is also how astronauts train. I have been able to turn my hobbies into learning experiences. I had the opportunity to teach extreme science when I became an aquanaut.

When I am not exploring underwater, I love to salsa dance and listen to music. I am also a cosplayer and have spoken at panels for Comic-Con. So, it is no surprise that I also love to read digital comics.

Above all, what fuels me the most is being around the people I love and being in a space where I can volunteer and give back. I give back because I know that you are the solution to the challenges of tomorrow. Therefore, I want to invest in you and am encouraged that you will play an instrumental role in our world's future.

The Connection

> *"I'd rather regret the risks that didn't work out than the chances I didn't take at all."*
>
> —***Simone Biles***, *Olympic Gold Medal Gymnast.*

When I began sharing my story, I was nervous and afraid of being judged or failing. Now, I am excited to share my journey with others, including my struggles and challenges, as well as my successes and highlights. I want to inspire others by showing that we all have to start somewhere, but the important part is where you finish. Be proud of every step you take on your journey.

I am excited to be on this adventure with you. Connect with me soon!

Connect with Sydney:

Instagram: @SecSydSoar
TikTok: @SeeSydSoar
Twitter: @SeeSydSoar
www.SeeSydSoar.com

GoZero took this photo while Sydney was on her commercial astronaut training experience with Uplift Aerospace's Space+5 Astronaut Class.

HOW A HIGH SCHOOL CLASS GAVE ME THE COURAGE TO MAJOR IN MECHANICAL ENGINEERING

Coauthored by Libby Brooks:
Project Engineer at Milwaukee Tool

My name is Libby, and I am a Milwaukee, Wisconsin-based project engineer for a company that develops and manufactures power tools. I graduated from Miami University in Oxford, Ohio, in 2019 with a degree in mechanical engineering and have been working in the industry as an engineer ever since. In addition to my career as an engineer, I run LibbyBontheLabel, an online business focused on career advice and content creation, and co-host *My Best Friend's an Engineer*, a weekly podcast for women in STEM (Science, Technology, Engineering, and Math).

While in college, I had two different professional work experiences: one as a Facilities and New Construction Engineer Co-op, and the other as an R&D Engineer Intern. Both experiences were at Briggs & Stratton, a company that manufactures small gas engines. I always get asked how I secured my first work experience with no previous professional engineering experience on my resume, and what my advice would be to college or university students looking to gain that experience. My best advice is to be yourself, and this is exactly how I believe I got my first co-op in my junior year of college.

I went to a small high school in the middle of Southern Wisconsin called Big Foot High School, where my graduating class had less than 100 people. Throughout high school, I was interested in all things related to art and took every art class that was available. In fact, I started repeating classes and taking them again since I was running out of art classes. In my last semester of my senior year of high school, I decided to try out a class that I knew absolutely nothing about. I figured this would be the best opportunity to learn about something that I otherwise would not have been exposed to. This mentality led me to sign up for a small engine class—this is where it all ties together.

I signed up for this small engine class and showed up for the first day of class in a pink velour Juicy Couture sweat-suit. I was the only girl in the class. The objective of the class was to learn about small engines, with the main focus of having hands-on experience. I vividly remember learning about the four cycles in a combustion engine: intake, compression, combustion, and exhaust. It seemed so intuitive to me and just made logical sense. The final project for this class was to bring in our family's push lawn mower, take apart the engine, put it back together, and ensure that the lawn mower correctly starts back up again. Sure enough, my family's lawn mower was held inside a Briggs & Stratton small engine, which I proceeded to successfully take apart and put back together. I later found that I was the only student in a class full of boys with a lawn mower that had successfully started again. This was the story that I told in my Briggs & Stratton co-op interview that landed me my first professional work experience.

My first job out of college was with a company called Toshiba America Energy Systems. At this company, I held the role of project manager and led the planning, initiating, and execution of refurbishment projects on power generation equipment. More specifically, I worked on steam turbines found at combined-cycle power plants. A combined-cycle power plant is one that utilizes both gas and steam to produce energy. The heat harvested from the gas system is transferred to the steam system. Steam is then forced through turbine blades, which cause them to rotate. The rotational force produces mechanical kinetic energy that gets converted to electrical energy through a generator rotor. My projects ranged from power plant inspections to services and repairs to retrofits. Needless to say, I know a lot of fun facts about power plants.

This was a great first job for me out of college because thermodynamics and heat transfer were two of my favorite courses in college to learn about as part of my engineering degree. This was one of the coolest jobs I could ever imagine having, and I would highly recommend anyone look into this field if they are interested in those types of courses. I was responsible for leading refurbishment projects from the quoting phase all the way through the final purchase order payment. In this role, I consulted with specialty engineers, managed client expectations, verified recommendations for repairs, worked with subcontractors, wrote and provided quotes and pricing for jobs, managed schedules, traveled to job sites, and the list goes on. I got to wear many different hats in this role and work with some pretty cool, larger-than-life equipment.

Other than professional experience, I grew up in a very small farm/tourist town in Wisconsin. I live near a lake that is a big tourist attraction in the summer, but during the three other seasons in the year the population is relatively low. As I previously mentioned, my high school was called Big Foot and had about 400 kids in it. I was one of maybe two kids to pursue engineering after graduation, that I knew about. Engineering was not really something I ever envisioned myself pursuing or really even knew about. No one in my family was an engineer that I could talk to, to get information about what the career path was like.

Why engineering? Well, after completing that small engines course I mentioned earlier, I had a burst of confidence and really thought I could take on the world. Knowing the high cost of college, I wanted to pursue an education that I truly wouldn't be able to gain anywhere else. To me, that meant I was going to become a lawyer, a doctor, or an engineer. In my mind, at the time, everything else seemed like something I could learn through life experiences and wasn't something I'd find value pursuing higher education for. Mechanical engineering it is! I truly decided to pursue mechanical engineering on a whim my senior year of high school after sparking a newfound interest in the small engines class, and I haven't looked back since. I chose mechanical engineering because my small engines teacher mentioned that there was some level of creativity and drawing found through 3D modeling. This interested me because I had focused my whole high school career on furthering my art skills, and still wanted to find a way to incorporate that into a technical role.

Today, I work for a company called Milwaukee Tool (MT). Milwaukee Tool is an American company that develops and manufactures power tools for heavy-duty contractors. If you were to ask me what were my top five companies that I wanted to work for out of college with my engineering degree, Milwaukee Tool would have been in the top three. I love that I get to work with young and ambitious engineers who are always looking to enhance the boundaries of engineering, to truly provide the most robust products to market. I work in the New Product Development Department on projects in the air movement group. That basically means that I get to work on any type of new-to-market tool that moves air. Think of vacuums, fans, air compressors, etc.

As a project engineer, it is my job to own and drive projects for the development of brand-new tools. MT prides itself on its speed and execution, delivering some of the top options to the market. Needless to say, it is a very fast-paced and exciting environment to work in. I manage multiple projects at once, with different team members and timelines, across different cross-functional teams within the organization. I get to work with all different types of engineers, including electrical engineers, mechanical engineers, design engineers, packaging engineers, motor engineers, and other project engineers. I get to work with suppliers, domestic and international, as well as domestic and international manufacturing facilities.

Qualities of a good project engineer include good communication skills, strategic organization of a project, time management, and attention to technical detail. I help with every aspect of creating a new tool, from creating prototypes and concept models to testing and evaluating its manufacturability. My favorite part of the role and this industry is the fact that I can go to my local Home Depot and see products that I helped bring to life sitting on the shelf.

Something that I am proud of myself for accomplishing within my engineering career, other than the fact that I earned an engineering degree and have made it as a full-time engineer, is that I passed the FE exam. The NCEES Fundamentals of Engineering (FE) exam is the first of the two exams engineers must pass in order to become licensed professional engineers (PE). Passing the FE exam grants you the title "engineer in training." This is a 110-question exam that essentially tests you on every single subject that you learned while getting your engineering degree. It is not required to pass in order to graduate, but some employers will require it as a prerequisite for a job. I have not encountered a job yet that requires this, but it does align with my values of continuous improvement and continuous education. I think it is important for an engineer to always be studying and learning about new concepts, which is why I am proud of accomplishing this feat.

Connect with Libby:

FROM CHEMICAL ENGINEER TO CHIEF WELLNESS ENGINEER A PATH FULL OF SURPRISES

Coauthored by Lennis Perez:
Chief Wellness Engineer at Just Lennis, LLC

I'm so glad you stopped by to read my story. I'm Lennis. I was born in Venezuela, and I spent most of my youth in that beautiful country in South America. Living there allowed me to visit the mountains and the beaches, plus I enjoyed plenty of family time. After a series of events, I ended up moving to the U.S. in the late '90s. My original intention was to spend a year in the U.S. to learn English, and then return to Venezuela to finish my studies in chemical engineering. Well, life had different plans, and I ended up staying in this country I now call home.

Engineering Just Makes Sense

I want to tell you a little bit about my own journey into engineering. Growing up in South America, specifically in Venezuela, it made sense to go for engineering. Why? You may ask. How does that career make sense? Well, if you know a little bit about Venezuela, it's an oil-rich country. And when you grow up in a country where you're driving by refineries and chemical plants, this becomes your normal environment. Also, my dad was a civil engineer, and when it was time for me to pick a career, my brother had already been attending engineering school. I also had cousins studying at different universities to become engineers, which meant I was often hearing conversations about being an engineer. It reminds me of stories that you may hear about families who follow the same career paths, such as doctors, lawyers, or even entrepreneurs. Once it's familiar in your environment, it becomes part of your comfort zone to stay in "the known" or "tried and true." At least that's how it felt for me. It was something that was comfortable because I was familiar with it. I knew I could get the support and help I would need as I went through school. I had the advantage of just tapping into this network of engineers in my family.

The funny part came when I decided to study chemical engineering. It was a decision influenced by two things:

1. I didn't want to go to the school my brother was attending, and that my dad had graduated from, because their program was extremely rigorous, plus they didn't offer chemical engineering.
2. And, during my senior year in high school, I was working with a tutor who was a chemical engineer and was extremely passionate about this profession.

I was also good at math; I understood chemistry, chemical reactions, and hard math equations (or so I thought), and during my senior year in high school, I took organic chemistry and fell in love with all the different carbon structures and how they related to everything in life.

But this is only the beginning of my journey. Fast forward to where I'm at today, and you'll learn I've taken a big pivot in my career. It honestly looks very different from anything I could've imagined when I was a freshman in college studying chemical engineering.

An Immigrant's Path

Before I get into how I ended up in my current role, I want to tell you a little bit about my trajectory as an engineer. I also want to encourage you to think about how your trajectory will be very unique to you. There may be some common things and themes, but ultimately your career, your path, and your life will evolve exactly how they were meant to happen for you.

As I shared earlier, I came to the U.S. to learn English. At the time, I knew being fluent in English would help me land a better job in Venezuela. This was especially true if I wanted to work for a big international oil company. Then, I learned very quickly that in the U.S. engineering school was actually easier than in Venezuela, believe it or not! So, I asked my parents: Would it be possible for me to stay and finish my career here?

During the same time, I applied for scholarships and was able to get a partial scholarship, so I could continue my undergraduate studies in the U.S. and graduate with my bachelor's in four years. Then comes the cherry on top: I was able to get a full ride to pursue my master's at the same school with just one additional year of classes. And that's how I obtained my Master's in Chemical Engineering from Manhattan College. However, as an immigrant, I had to face the challenging task of finding a sponsor so I could stay legally in the U.S. Fast forward to the spring of 2005. I was so happy that I was able to find a company that offered to sponsor me and landed a job where I got to design chemical plants. This was the closest I could get to my dream career as a young engineer. My original dream was to work for an oil company or at a refinery, but that actually didn't happen.

The Dream Wasn't Real

Within the first five years of working as an engineer, I was learning, growing, and had a "short-term" path figured out. I was part of a rotational program within the organization where I would work in different departments, and once I completed that program, I would take a permanent position as a process engineer for one of the departments. I chose the department that worked on designing refineries, trying to get as close to my dream as possible. Then I started to feel something was off. I started thinking about what was next in my future. Although it was still early in my career, I could see two distinct paths I could take. I could go into sales and management, or I could become a technical expert. But inside there was a voice that would whisper, *This place isn't for you.*

Looking back, that voice showed up when I became aware of my role model's relationship with work. When I looked at my managers and other leaders in the organization, I felt they just *did their job*. This inner whisper told me I wasn't going to have as big an impact as I was starting to realize I wanted to have. And with that, I ended up switching companies. An old colleague, who had been laid off, introduced me to the new place where she was working. This was a small consulting and design firm, almost like a mom-and-pop shop, where the owners were the principal engineers. Without having full clarity on what my real dream was about, I accepted a position as a process engineer at this smaller firm. I was excited to see that the process engineering department had many women engineers working there. The split was close to 50/50.

Turbulent Waters

About ten years into my professional career, I started navigating some personal challenges. Then, in early 2016, I got laid off. This was a painful experience, but because of all the other personal stuff I had been dealing with in the past few years, I had started to recognize that my job title was not my worth. I was thankful that I had set myself up in a financial position where I could take some time off. I then decided to take a career break. This is a term I had never heard of before or seen anyone use. I had seen women take time off because of maternity leave, but never because they needed time to recalibrate their lives. During my career break, I ended up helping my parents with some things here in the U.S. My mom moved in with me after the political and social situation in Venezuela created an environment that was hurting her emotional and mental health. Five months into my career break, I got a call back from the company that laid me off with an offer to work for a different department, and I accepted this new role.

Looking back at that rough period in my life, I can identify one of the challenges I was facing as not knowing how to ask for what I needed. I lacked the tools to communicate assertively and to share my frustrations while also creating a plan of what we (my manager and I) could do as a team. I was struggling with one of the managers, as I felt he was micromanaging my responsibilities. I didn't know how to set boundaries, and I had been going through a lot of other stress. At the same time, I started to see how many of my coworkers and colleagues were falling ill. Many were dealing with burnout symptoms and other health issues, which I started to recognize in myself as well. That was really one of the most challenging times in my career and in my personal life: Trying to navigate through burnout and how to deal with stress, as well as how to deal with micromanagers and how to set boundaries. All these things that I wish someone had taught us in school.

The Light at The End of the Tunnel

After recognizing I couldn't manage the situation at work (because I didn't know how), I decided to leave. As I was making that decision, another job opportunity came up. I'll say this new job felt like one of my greatest accomplishments in my professional career. I became the youngest and only female Latina engineer on the team. I was given the responsibility to manage the expansion of the process systems product line, bringing these technologies to the U.S. and the Americas from Europe. This was a huge undertaking, and they trusted me with being responsible for this entire continent (North, Central, and South America). I was able to travel to Brazil, Argentina, Germany, and other parts of the U.S. territory. It was exciting, and I felt like I was living my dream life, one I hadn't dared to dream before.

But yet again, a couple of years passed by, and here I was, reliving that "off feeling" that hunted me multiple times throughout my professional career. And now I can tell you with certainty, the biggest challenge I faced in my career was *not being in alignment with my truth*. And when you feel out of alignment, when you feel you can't bring your whole self into your career, it sometimes feels as if the walls are closing in on you. It took courage to accept that it was time to leave and to find my authenticity and the place where I could truly belong. This journey led me to become the chief wellness engineer for Just Lennis. I realized I didn't fit in, and I didn't want to fit in anymore. I ventured into entrepreneurship and created an organization that supports professionals in STEM (Science,

Technology, Engineering, and Math) who are struggling and who feel they're alone in their journcy.

Removing the Veil

When I was going to school and during the early years of my career, many of my challenges were issues that were handled by going to therapy. Something I did when I turned 29 years old. In my experience, nobody talked openly about how they navigated personal or professional challenges. These issues are "yours" to figure out. At least that was the work culture. And for years, I wanted to hear others speak openly about what it's like to be a woman working in a male-dominated field. I wish someone spoke openly about how to handle microaggressions or toxic work environments. But these are issues we don't talk about. So here I was, again, trying to figure out what to do, except this time it was different.

I had been working for almost a decade on gathering tools that would strengthen not only my intellectual growth but also increase my emotional intelligence and spiritual well-being. I'm inspired to work with fellow engineers and STEM professionals because I see the value they bring to the world. I see the potential and positive impact we could have for our fellow humans and the planet in general. And it shouldn't come at the price of sacrificing their physical health and mental well-being. Working as a chief wellness engineer with professionals in STEM, and creating spaces within organizations to have sustainable strategies and tools to help team members manage stress, allows individuals to not only enjoy work but also show up as the most creative and healthy version of themselves. I believe this leads to creating the best products, services, strategies, and solutions to continue improving the world.

Final Thoughts

As we continue to promote more diversity in engineering and STEM careers, I invite you to reflect on the tools you can provide to help these young minds equip themselves properly to navigate the challenges they'll face and to know they won't be alone.

If you, your team, or your organization would like to continue building up a sense of belonging amongst the individuals involved, let's work together on creating these spaces to move the needle forward. I'm here to support you in any way I can.

Thank you for reading. I'm really excited to be part of this wonderful book, and thank you, Dr. Justina Sanchez, for putting these inspiring stories together.

Connect with Lennis:

STEM WITH JEN

Coauthored by Jen Patel:
Senior Engineer at Becton Dickinson (BD)

Flashing down the hall, the little girl and I rushed to the emergency room. She worriedly yelled, "Hurry! Hurry! Her leg is tearing!" As we gushed inside, I kneeled down and meticulously took the injured patient from the little girl's hands. After suturing in a rainbow patch at the tear, we exclaimed, "Operation success!" The little girl and I jumped, zealously ecstatic over the surgery. The pink-and-blue-polka-dot teddy bear was saved.

This was one of the moments when I realized my passion to help bring a "sigh of relief" to others. Growing up as an active volunteer at my local children's hospital, I witnessed how advanced medical technologies profoundly impacted pediatric patients' lives, but I also recognized the challenges that existed in diagnosing and alleviating many of these ailments. Being with these patients and families daily and seeing the challenges they were experiencing was the catalyst for me to build a career to improve their lives. I aspired to be a significant contributor to medical research and technology in hopes of bringing a "sigh of relief" to patients and their families.

I am a first-generation biomedical doctoral student, mechanical engineer, and science communicator. I obtained my Ph.D. in Biomedical Engineering and Bachelor's of Science in Mechanical Engineering at the University of Tennessee at Knoxville. I grew up in the mountain resort city of Pigeon Forge, Tennessee, as the daughter of immigrants and did not always aspire to be an engineer.

In fact, all through grade school, I aspired to be a pediatrician. I never considered engineering for two reasons. One, I was exposed to engineering as a mundane, male-dominated profession where I saw no one like myself. Two, I was not exposed to the diversity of engineering other than seeing the mechanics work with car engines, which at that time did not interest me. It was not until my final year in high school that I shadowed a pediatric surgeon and was intrigued by all the tools used at the hospital. He suggested that I look into biomedical engineering. I was intrigued by the field and saw it as a field where I could apply my passion to bring a "sigh of relief."

Soon after, I applied to the University of Tennessee and was accepted into the Biomedical Engineering Program. But little did I know, *there were so many engineering disciplines!* As I learned more about all the fields, I realized I loved researching, designing, building, and 3D modeling. I switched my major to mechanical engineering to take courses that better fit my interests, and I saw that mechanical engineers also work to advance health. Anyone can be an engineer, and there's more to being an engineer than the stereotypically dull job description. There was more to even being a mechanical engineer beyond designing car engines (even though it was cool learning how engines work as an undergraduate). Engineering involves so much more creativity and diversity.

However, pursuing a career in engineering is challenging. There were so many times I wanted to just quit. Balancing coursework and understanding engineering with a little physics background was hard. I failed my first few exams and so many others throughout my undergraduate studies. I applied to nearly a hundred research positions and internships, and faced so many rejections. My time management fell through the cracks, and I did not know how to make time to spend with friends and family. The worst of them were often told, "Sweetie, you do not look like an engineer."

But there was one thing I was learning and did not realize at the time—finding ways to be perseverant and resilient. I was consistently working through the challenges and building a supportive community. I joined the Society of Women Engineers (SWE) section my first year, joined the executive board the following year, and eventually became president my last year in college. I commemorate such an organization for providing resources, experiences, and an extensive network that allowed me to find the inspiration to "be that engineer." Through this organization, I was also able to find everlasting friendships and a supportive community that helped me be perseverant and resilient.

I also found the current SWE mentoring program at my university. My first year, I realized through academic and career challenges that my peers and I were facing a significant lack of guidance. I sought help to better learn about careers that fit me but had difficulty finding mentorship. After my first year, I wanted to provide opportunities for future first-year students to have access to mentorship for academic and career questions. Thus, I founded a professional development goal-oriented mentoring program with the help of the SWE section, where upperclassmen were paired with first-year students and given semester academic and career goals. The program resulted in over 120+ members and 1200+ mentoring events reported and was honored with multiple awards through the years at national SWE conferences.

I now serve as an alumni mentor for the program, where I encourage my mentees to take the initiative to address challenges they see. Take the initiative to build your own community. Be the change agent and take the initiative to build not only your own pathway, but a pathway for others as well.

As graduation was approaching, I was anxious and confused about building my future career pathway. I interned at Siemens Healthineers as a mechanical engineer during my undergraduate years and loved the experience and mentorship I received. However, I still felt I wanted to learn more, but I was not sure how to make this happen. I met with one of my biomedical engineering professors, hoping to receive career guidance, and was challenged by the professor to determine my interests. I reflected on my past experiences, including volunteering at the children's hospital, and realized I wanted to learn more about being an engineer in pediatric health. My professor suggested I research Ph.D. programs where I could continue my education and train to be a researcher in the pediatric field.

Little did I know, my mentor had just received a tenure-track promotion and was seeking to be a graduate research assistant to advance drug delivery research for pediatric cancer. I expressed interest and applied to be the first graduate research assistant for the laboratory. In the few months that followed, I also explored other programs at other universities, hoping to find other interests in the pediatric field. I was offered positions at other laboratories; however, my final choice came down to choosing the position my mentor offered because I prioritized my mentor-mentee relationship with the principal investigator (PI), and I would be able to learn and apply practical knowledge to advance the pediatric field.

A month after I graduated, I started graduate school to pursue a Ph.D. in biomedical engineering. My Ph.D. research primarily focused on innovating and evaluating an injectable hydrogel-based local drug delivery system designed with hydrophobic and hydrophilic chemotherapies to treat pediatric brain tumors and other cancers. As the first and only graduate research assistant in my laboratory, I independently initiated, developed, and executed protocols for material characterization, drug release profiling, in vitro cytocompatibility evaluation, and in vivo tumor tissue response.

Just like many graduate students, I was excited and motivated to start my Ph.D. research, but I soon realized pursuing a doctoral degree is a marathon. I felt like I did not know anything, and all my experiments were failing. Once again, I wanted to quit.

However, the challenges I was facing this time were exponentially elevated, as opposed to when I was an undergraduate. The coursework and research were difficult to understand. As the only graduate research assistant, it became my role to manage the daily laboratory activities and the undergraduate assistants. I had trouble managing my time and was not even able to make time to spend with friends and family. Similar to when I was an undergraduate, I sought mentors but did not know how to find them.

As a result, I founded the professional development mentoring program for women in STEM (Science, Technology, Engineering, and Math) pursuing a doctoral degree. After months of planning, I launched the program with over fifty women in STEM present. Through this program, I was able to find a postdoctoral mentor. This mentor, along with the help of my PI, was able to guide me in developing a more feasible action plan for balancing research, coursework, and social life. Networking and a supportive community can have a profoundly positive impact on pursuing a Ph.D.

An unexpected challenge during graduate school was the pandemic—a time of unknowns. Graduate students were sent home and were no longer allowed to continue their laboratory work. Research from previous years seemed to be no longer applicable to obtaining a Ph.D. Upon return, supplies of essential laboratory resources, including personal protective equipment like gloves, were slim and priced beyond allocated budgets. Ph.D. graduation was unforeseeably delayed.

Arguably one of the most important decision factors in selecting a Ph.D. program is selecting the PI to build a mentor-mentee relationship with. This was a time I was grateful to work one-on-one with my PI to brainstorm and navigate the challenges of the pandemic. I vividly remember a time when all shipments kept getting lost, so I drove twelve hours over a span of two days to receive a package crucial to my research. Following this trip I and graduated with a doctorate in biomedical engineering. If there are any skills to be learned during graduate school, one is the continuous cycle of being perseverant and resilient—finding pathways even when it seems impossible.

I am now on a mission to engineer future medical innovations, promote STEM education, and empower future engineers. I work as a senior engineer at Becton Dickinson as part of the Technology Leadership Development Program, a highly selective program aimed at training the future leaders in the medical device industry.

In the STEM community, I am known for my STEM influencer brand, STEM with Jen. I founded this brand upon my Ph.D. graduation to engineer, educate, and empower a supportive STEM community. Being a first-generation engineer, I did not have a strong STEM network growing up. In fact, I did not even know what biomedical engineering was until I started pursuing my undergraduate degree.

Over the years, I learned to build my community of role models, mentors, advisors, STEM friends, etc. But knowing others, especially younger students, do not have access to a STEM community, gives me fuel to build one for them. I want to expand the STEM community and make engineering a field accessible to everyone. Follow along as I share my adventures as an engineer through @stemwithjen.

I urge you to take the journey that molds you to be resilient and perseverant. Don't be afraid to take the next step or even dream big; just go for it! Make mistakes and learn from them. One of my favorite quotes is:

> *"Shoot for the moon. Even if you miss, you'll land among the stars."*
>
> —Norman Vincent Peale

Connect with Jen:

THE PEOPLE ENGINEER®

Coauthored by Brianne Martin:
People, Process, & Performance Consultant

Hi there, Brianne here, also known as The People Engineer®. Before we dig into the nerdiness of my love for engineering, please allow me to introduce myself. I'm a high-energy extrovert who is not only obsessed with all things math and science in the world, but also with our most untapped resource—humans. As a fourth-generation Tejana (Texan), I am a Mestizo: a mix of Native American, Spanish, and Mexican. I was raised by a single mother. I'm the oldest of three, and my family loves sports. We were all athletes growing up. I started school at ten months old, thanks to a Methodist Day School, and I've been hooked on learning ever since. Starting in elementary, I was able to attend a Montessori school, which encouraged my hands-on learning. By middle school, I was in honors math and science classes. You see, I've always loved math and science, but they stopped loving me back around eighth grade—algebra and physics.

Random fun facts about me: I love singing and have been doing so on public stages and in front of large crowds since I was three years old. Rainy days are my favorite. My biggest smiles happen on a dance floor; cumbia, two-step, salsa, Tejano, you name it! Singing, dancing, and building stuff with my hands are my deep, soul-fulfilling hobbies.

Growing up, I was amazed at the world around me. To this day, playing with leaves and observing plants and animals is astonishing. Don't even get me started on the stars and universes. I remember first learning about photosynthesis and how plants grow toward the light. My mind was blown! As if it were yesterday, I recall being in my fourth-grade class and being very inquisitive about the new chairs in the classroom. I had only been around wooden chairs for the majority of my school career, but THIS CHAIR was plastic with shiny metal legs and had multiple silver buttons. It was wild to me that this shiny metal was bent. How did it know not to bend too far? How heavy could I be before it broke? Could I jump on it? How many times? What if the plastic cracked? This moment began my heightened interest in manufacturing. Although I didn't know it at the time, my longing for answers and genuine curiosity for the world around me would lead me to be an amazing engineer.

If we back up a couple of generations, my grandpa (or Popo, as I called him) was the first person I knew who had a knack for STEM (Science, Technology, Engineering, and Math). I remember going over to his house, and he would have these massive scrolls of blue paper. He would unravel them, taking up the whole table, and show me how the paper was blue. The drawings would depict numbers, lines, and descriptions of some type of building, home, or even a room. I was fascinated. Not only had I never seen blue paper with white lettering, but there was also the fact that there was a two-dimensional picture of something that my Popo was going to make in real life. It was like color-by-numbers instructions on how to do something. I didn't know that's how building stuff worked. Ironically, my Popo had never gone to school. He had been working hard, doing manual labor since he was five years old; he never learned how to read or write his name, but here he was building massive schools, libraries, homes, movie theaters, and more! Calculating things came naturally to him. At a glance, he could size things up. He was smart, funny, and could make sense of the world. My Popo had superpowers.

The saying "the apple doesn't fall far from the tree" was very fitting with my mom being my Popo's daughter. She, too, seemed to have a knack for math and science. My mom has pictures of herself in high school during her tech class, where she shaped metal into the outline of a horse. There's also a picture of her and a barbecue pit she welded herself. It was pretty cool to know my mom was in marching band, played the clarinet, was the lead twirler, played basketball, and was able to build things with her own hands in high school. However, back in the '80s, when she was in high school, there was still the false belief that young girls were not suited for STEM (Science, Technology, Engineering, and Math) fields. So, despite my mom clearly having both talent and skills in her math and science classes, when she approached her career counselors at school about her future, potential, and college plans, they directed her to accounting as her ideal career path.

Through these two generations of mechanically inclined family members, my curiosity as a child was not only encouraged but elevated. While the majority of my education happened in the great state of Texas, during my freshman year in high school, my family unit of four moved to Elko, Nevada. To my surprise, when registering for high school, the Elko School District provided me with a catalog of electives to consider, including classes like auto, shop, woodshop, and drafting. I was so excited to explore, especially because in Texas, the only electives were choir, band, and art. Initially, I had the hardest time trying to choose between auto shop, and woodshop. I wanted to build stuff. I wanted to take things apart. I wanted to learn these supercool skills. However, my mom, knowing what she knew at this point, encouraged me to take drafting. She said, "You can always have auto, shop, or wood projects on your own, but drafting is a strong skillset. If you learn how to draft, you can design and create plans for future projects for shop, auto, and woodworking." As disappointed as I was in the moment, I trusted my mom and registered for drafting.

The first week of my drafting class seemed a bit anticlimactic. We learned how to tape down our paper, the different thicknesses of pencil lead, and the different tools we'd be using: a T-square, circle templates, triangles, and scales. I thought, *Oh how boring*. However, once we got into the swing of things, my experience and attitude completely changed. You see, stick figures are about my highest level of my artistic ability. My little sister was the artist in the family, and she was amazing. So, when I tell you how cool it was to be able to draw a 3D isometric view of an object and actually make out what it was and understand it, this made me feel a great sense of pride and accomplishment. Hand-instrument drafting has become a lost art. It's both art and science, but through the boom of technology, like most things, it now lives in a software program widely known as CAD (computer-aided design).

In the second semester of my freshman year, I had my first CAD lesson. While I was heartbroken that we were no longer using our drafting tools, it was pretty neat to be able to gain speed in designing, learning commands, and shortcuts, and even adding animations to our designs. At one point I asked my teacher, Mr. Aikenhead (yes, that was his real name), "What kind of job can I do to use these types of skills?" His response was, "Architecture or Engineering." And thus began my infatuation with engineering.

By the time I was a senior in high school, I had already attended three different high schools. Life had an odd way of ensuring I was adaptable, and through sports, I'm grateful I was always able to make friends. When it came time to make a decision about college, I knew I was going into mechanical engineering. Remember my excitement about Auto and Shop? From my understanding at this point, mechanical engineering was learning how to not only understand how machines work, but also how to design, improve, and make things in those areas better, faster, and stronger by using the skills I learned in drafting. It was the best of both worlds!

When looking for schools, I would cross-reference universities that had my degree in mechanical engineering, and that of my friends, through a tool called College Board. Most of my friends wanted to stay in Texas, but others were eager to get away from their small towns and go across the country. I had done research at school and fell in love with a few engineering programs just based on their websites. Notre Dame was my top choice. I envisioned myself in the Midwest, enjoying the seasons, football games, and the family environment I had always seen promoted; and let's not forget that as a sports family, *Rudy* was one of our favorite movies.

When applying to different universities, it made sense to have backup schools and keep an eye out for other options, and it just so happened I stumbled upon Marquette University. For whatever reason, the pictures spoke to me. The school's slogan: *"Be the Difference."* Like Notre Dame, it was a private university, but everything I read focused on service. We connected virtually right away. It was difficult for me to have these strong feelings, aspirations, and visions of my future while facing the reality of not coming from a wealthy family. Granted, I'm always amazed at how my mom worked so hard and strategically that my brother, sister, and I never went without, but college was expensive, especially at these private schools and being out of state.

I'll never forget, at eighteen years old, around the time of my birthday, after submitting all my college applications, my mom and I sat at our kitchen table in our apartment for "the talk." My mom expressed how proud she was of me, and said that this was a big step for me. She explained loans, finances, and how she was willing to help me by taking out a parent loan, but that I had to stick with my initial major choice. "Brianne, some people will have the luxury of being able to change their minds. We do not have that as an option. It's already going to be a stretch financially, but if this is what you want, you'll have to dig deep to make it happen. I know you can, but please understand what a big decision this is. If you choose mechanical engineering, this is what you'll get your degree in. If you change your mind, don't like it, or want to do something else, you will get this degree, and then you can go back or find a way to change your career path as a mechanical engineer." It felt permanent. It felt heavy. It was both terrifying and wild to know I could make this big of a decision for myself—my life. We agreed. We even shook hands and laughed. She hugged me, and we both got a little emotional. I couldn't see myself doing anything else. This was my path.

College acceptances started as early as the holiday season in December, especially for my fellow sports brothers and sisters. People were committing to junior colleges, four-year universities, and even the Ivy League. The spring is when everyone is already sure about where they are going, and I am still waiting to hear back from my college. That big conversation between me and my mom was due to the grueling financial aid paperwork, scholarships, and submissions to the schools about how I could come up with the money for classes, room and board, and fees if I were admitted.

Now, I was getting ready for mid-terms, final projects, and only a few months away from graduation. I was checking our P.O. box daily. I got the state university acceptances first. These were my backup schools, just in case. I felt a bit of relief, a little excitement: "Okay, I have options," but I was holding out for Marquette. The letter finally came. On Marquette letterhead. My friends had told me to be on the lookout for a big packet. Most universities would send the course catalog for you to choose the first semester of classes along with your acceptance letter. I received a small letter. My heart dropped. I prayed, "God, let this be them letting me know my big packet is on its way." I told myself, *Maybe they just do things differently. This isn't bad; this is going to be good. Maybe they sent me school stickers to show my Marquette pride*. I opened it. It was just one piece of paper. "Thank you for applying. We cannot admit you into the College of Engineering at this time. You've been added to our waitlist." I felt sick. I HAD to go to Marquette. I was meant to go there. We were made for each other. I felt so many emotions at once. I was mad, frustrated, defeated, enraged, heartbroken, devastated, hurt, not good enough, and disappointed.

After a hard month of trying to figure out my options and starting a couple of summer classes at the local community college, I finally got the letter of acceptance and my course catalog. Naturally, I still have this course catalog on my bookshelf to this day. The mental work began. You see, Marquette University was 1,149 miles away from home. I was grateful my mom was able to get a great deal on my previously owned but basically new car earlier that year, but that was a long way away from my family. Thankfully, my mom has always stressed the importance of independence. As a single mother, she wanted to ensure my siblings and I could always take care of ourselves if anything ever happened to her. Small things like making us order pizza over the phone, filling out our own medical forms at the doctor or dentist, and even giving us cash and dropping us off at the grocery store to pick up milk and bread. Realistically, I knew I would be okay, but this fear of being so far away from home made my stomach turn. Because of the quick timeline of how this all unraveled, I did not have much time to think about my fears; it was go time! I was constantly applying for scholarships. I had finally gotten the chance to start working part-time (my mom never allowed it before), and I was planning everything I needed for my dorm, the seventeen-hour road trip, and my classes for the fall 2008 semester.

Fast forward to being on campus, sitting in the front row of most of my classes, and truly living life. It was easy to get into the rhythm of things. I made friends easily; I lived on the all-girl engineering floor of our dorm, and I even found a few friend groups to socialize with and go on random adventures with. School was always my priority. I was the nerd who would stay in my dorm on Friday and Saturday nights. If a group of friends made dinner plans, I'd go out, enjoy myself, and then come home to crank out 3D modeling, chemistry, or dynamics homework while everyone else continued to enjoy their nights at house parties or dorm gatherings.

Despite my dedication to my studies, I was having a really hard time with calculus II. Calculus I gave me a run for my money, but I passed by the skin of my teeth. While calculus I is differential calculus, calculus II is integral calculus—essentially a tier two of revisiting all the mathematics of calculus I, just going backward or "the other way," and introducing calculus III fundamentals, which add in higher dimensions. (For anyone interested, calculus II covers integration, differential equations, sequences and series, parametric equations, and polar coordinates. Calculus III covers parametric equations and polar coordinates, vectors, functions of several variables, multiple integrations, and second-order differential equations.) So here I was putting in as much time as humanly possible to study, going to tutoring, visiting my professor during office hours, keeping a part-time job, and making time to both eat and sleep, and it just was not cutting it.

Inevitably, I failed calculus II, and not just once but twice. In the end, I registered for calculus II five times, failed twice, withdrew twice, and finally passed the fifth time with a B. But let me back up.

The first time I failed, I was learning it from the math department chair. He was an amazing professor, told great stories, made the material digestible, and genuinely had a great time teaching, so his energy made the class fun. The second and third times I withdrew, I had a not-so-great teacher. The fourth time, I had the pleasure of trying again with my amazing professor. He was aware of my struggle, and we were both determined to have me pass this time. As you can imagine, this was my fifth semester in school (having registered three times prior). So, my fourth attempt was in fall 2010. The hard part was not just struggling in the class; my pride took a beating. My confidence was shot after failing once and withdrawing once. I called my mom, crying. "Maybe I'm just not smart enough to be an engineer. This is an entry-level class, and I'm stuck taking it over and over. I'm not getting any better. If anything, it feels like I'm getting worse. I know I promised I wouldn't change my major, Mom, but I can't do this."

My mom, in all her wisdom, consoled me. "Brianne, who cares how long it takes you? If you're serious about changing your major, what would you want to do in life if you're not going to be an engineer?" I replied, "I don't know. A teacher? I've always liked teaching, but golly, I just really want to be an engineer. I love everything I'm learning in all my engineering classes. It's interesting and it all makes sense. I'm truly passionate about it all." I replied, "Okay then. If it takes you ten years to finish your degree, who cares? You want to be an engineer, right? You have the rest of your life to work. A few years will be a drop in the bucket. If you're really doing all you can, maybe try another approach. Find a different teacher, a different school, and if you simply need to, take a break and then try again—don't give up."

That conversation still replays in my head today. I'm incredibly grateful and blessed that my mom believed in me. She could have scolded me, reminded me how much money was on the line with not only my personal loans but the parent loan she took on herself. She could have coddled me and told me to come home. But she knew better. She pushed me. She let me cry it out and be frustrated with the situation, and then lifted me up, dusted me off, and put me back out there.

By the end of my sophomore year at Marquette, I was beginning to take more in-depth engineering classes. I had aced all my engineering classes, but my "core classes," calculus and chemistry, had hurt my GPA. I was put on academic probation. This is essentially the university saying, "Hey, Brianne—you're at risk, and we're watching you." It's not a good feeling. Knowing I was on academic probation added more stress to my already loose footing. Furthermore, I was having a hard time concentrating in class. I took notes like crazy because I could not for the life of me recall what was said in class. It got so bad that I even started showing up to class with a tape recorder. I'd have to go over my notes and listen to the recording several times over, and I had to do this for all my classes. I noticed this became a more serious issue throughout my sophomore year, and thanks to the university's resources, I started seeing a counselor on campus.

First, it started with a few therapy sessions, learning techniques on ways to calm my anxiety and tools to better manage my stress, but with the increase in my inability to focus on class or my starting to have panic attacks during quizzes and tests, I was tested for and diagnosed with ADHD. There were also a lot of conversations about my childhood, upbringing, and experiences. While I was in high school, in my second year (from sophomore to junior year), I had seen a CASA (Court Appointed Special Advocate) volunteer and counselor regularly due to having intense mood swings and out-of-character outbursts. As a teenager, I knew hormones and mood swings were a thing, but I would scare myself with the high shift in emotion and my state of mind, from depression to manic and back again. I was raised to be in tune with myself, my emotions, and state of mind, especially because I experienced various levels of abuse as a child. It's still something I'm processing and working through, but as research shows, our mind and body take over and rewire themselves to survive through any major trauma.

So, tying this back to my college studies and not being able to focus in class, my anxiety and decreasing levels of concentration came to a peak, and thus academic probation only put it in writing. I'm a rather spiritual person, and I firmly believe God put amazing people in my life to serve as a blessing. Dr. Krenz (my first amazing professor) was one of those blessings.

He was well aware of my struggles in class, and after my fourth attempt, he scheduled time with me to go through and troubleshoot exactly where I was getting held up. He and I met in the math building, on his personal time. Problem after problem, he wrote out an equation and watched me as I set up the problem. I wrote out my thoughts until I got to a point where my brain would fog up and it would just stop. He looked at me and smiled while shaking his head about three problems in. "Brianne, I have great news and not-so-good news. The great news is you get calculus II. You understand the concepts, and you do a good job of showing your work and setting up the problems. The not-so-great news that is you lack basic skills in algebra. Solving for X is where you begin to fumble." This was a revelation. All that I had learned and researched about people with PTSD (post-traumatic stress disorder) was that the brain will suppress specific times in your life when abuse or trauma occurred to protect itself. So, the good news was that it wasn't that I wasn't smart enough; my brain just wasn't willing to go back and recall previous things I had learned because of the timing of when it all happened.

I wish I could tell you that after learning this crucial explanation of what and why I was struggling that I found the solution, unlocked the access to my eighth-grade brain, and was able to overcome this embarrassing moment in my life. But alas, this is not a fairy tale, and my reality was much harsher than that. I was put on academic probation again—for a second time—and then asked to leave my dream school, Marquette University. To put it plainly, I got kicked out.

The cool part about engineering is that every summer there are opportunities to work for companies as an engineer to learn how to be an intern, and the company gets a chance to groom you and have basic engineering tasks handled at a much lower rate. Through my tenacity, willingness to learn, positive attitude, and outgoing personality, I had been able to secure an engineering internship every summer since my freshman year in college.

My first internship was for a city where I served as a civil engineering intern. It was not aligned with my degree as a mechanical engineer, but I figured beggars can't be choosers, and it was rare for a freshman to get any kind of internship. So, I would show up, do my best, and keep an open mind to look for transferrable skills. My second and third internships were for the United States Army, supporting helicopter repairs. Through my work, I was able to design a 3D model of an aircraft using AutoDesk Investor and referencing physical blueprints, update and revise maintenance engineering orders, develop standard operating procedures for overseas soldiers, and support various projects and daily tasks supporting aircraft engineers and mechanics.

Thanks to this amazing network and mentors at the army, when I shared the news of my school struggles, I was reassured that engineering degrees are the hardest to earn and aren't for the faint of heart. Once I was asked to leave Marquette, my previous bosses and intern mentors advised me to move close to the base I had previously interned at. At the very least, they were impressed with my work and could get me a part-time job while I figured out my next step.

Fortunately, the head of our division told me he served on the industry board for the local state university, and they were introducing a new mechanical engineering program. He advised me to transfer to this college, take basic classes to boost my GPA, and then transfer into their engineering department to finish my degree. Thanks to these amazing souls and their advice, that's exactly what I did. It did feel a bit like I was punishing myself for failing by having to move back to

Texas and finish my degree at a state school, but I'm grateful I had a soft landing and a job to fall back on.

The uphill climb continued to get steeper and steeper with each step. Yes, the state school was nowhere near my dream school, but most of the professors were okay. The dean of engineering and the staff in particular were very helpful and kind. Honestly, a few of my professors were the exact opposite of kind. Whereas, most of the engineering professors I had at Marquette loved their jobs, enjoyed teaching, and had a blast leading their classes, some of my professors at Texas A&M University–Corpus Christi made it seem like we were burdening them. Some even had the audacity to tell me I would "never make it." One, in particular, told me it was "cute" that I thought I could be there and that I was even trying engineering.

This behavior only worsened once I continued my therapy utilizing the university's resources, got established with a therapist, and finally had the chance to begin medicine for a confirmed second diagnosis of ADHD.

My senior year, I ended up testing for twelve different medications. Some made me super hyper; some turned me into a zombie; some had no effect on me at all; and some made me so sleepy and groggy it was painful to fight to stay awake in class. Thanks to the disability services the university had, I was provided resources for individual quiet rooms to take quizzes and tests, but I had to get my professors to sign agreements to utilize these resources. I was heartbroken when one of my professors I got along with laughed in my face and told me it was "unfair" and "only in my head." He wasn't the only one who had this type of attitude, because back in 2012 mental health and psychological disorders were not being discussed or even understood as much in the education world as they are now.

Even though it is expected that life has its highs and lows, my engineering career has continued to surprise me. Not only in making it through to earn my degree, fighting through deep depression, having major family drama blow up during finals week, failing classes, and so on. I knew I was a great engineer once I had a role. When I was in engineering, I had a blast. My soul was on fire. Working in teams, providing solutions, and being creative was so much fun. I loved to learn, I loved to serve, and I really enjoyed the world of engineering.

In 2013, my super senior year, I attended my very first SHPE Conference. At the time, SHPE stood for the Society of Hispanic Professional Engineers. They have since shifted to supporting all Hispanics in STEM. But this was my last semester in school. I had already secured a full-time position from my internship the summer before, so it was pretty amazing to not feel the pressure of having to find a job. What I did find were mentors, role models, and people who were not only dreaming big but living it!

I'll never forget this workshop as a part of the SHPEtinas track (a track dedicated to Latinas in STEM) where a Global V.P. of Operations was the speaker. She was a Latina. She shared her journey of changing jobs, moving around the country, and sometimes having to shift roles for better pay or better opportunities. She said, "Your career is more like a jungle gym than a ladder." As she shared her personal journey, she highly encouraged everyone to consider rotational programs. This is where companies acknowledged the value of early career engineers understanding various aspects of the business, providing them with an outline or roadmap of the various departments, plants, or even locations, and then raising them to a higher level of management once they complete it. This was music to my ears. That was exactly what I was looking for. Granted, I knew I didn't have the GPA to get into these rotational programs, and since I already had a full-time position, I was grateful for what I had, but now I had the vision of what I wanted to do.

I began developing my own rotational program at an international aerospace manufacturing company. I started as a Tooling Engineer, then moved to the Liaison Engineering group. As a tooling engineer, I designed over 30 new tooling concepts, resulting in an increase in safety and productivity, and a reduction in manufacturing costs. I also oversaw the first 3D printer for the company (a $500k piece of equipment). Once I moved to the Liaison group, I was seen as an asset and moved from Assembly Line Liaison Engineer to Liaison Design Engineer to support the aftermarket sales team, modification kits, and service request updates. My thorough experience in design allowed me to bounce between the design world and the manufacturing world.

Even though I had realized that design engineering was no longer for me, I utilized the time and experience to better understand the aftermarket sales team I was supporting. At the time, there were only three young men, all with business or marketing backgrounds, who were traveling the world to and from customer sites. They'd bring me back a list of questions, concerns from customers, and a bunch of ideas, along with amazing photos from their international travels. I immediately thought about how cool it would be for *me* to travel to all these exotic places, sit face-to-face with customers, and help them come up with solutions on the spot! I knew how we make all of our parts. I would know what changes would cost them a fortune and what small adjustments we could make to appease their wishes, only requiring tooling updates rather than an entirely new design. Bold idea initiated.

One day, during an impromptu assembly line escalation, all the bigwigs of the company came to my GEMBA walk to review the current status of this big order that had a critical problem—a big deal, a big client, costing us hundreds of thousands of dollars every hour. The V.P. of Marketing and Sales was impressed with my initiative, attitude, and leadership in taking the next steps to provide solutions to the problem at hand. He pulled me aside and said, "Your energy is infectious. Keep up the great attitude. It makes a difference around here." His words still stay with me today. A week later, I built up the courage to walk up to the executive officer and ask for ten minutes of his time. He asked me to sit down, and I pitched my idea of being his version of a sales engineer. I had learned about these types of roles through my experience with SHPE, that companies had these positions, and the job descriptions sounded just like me on paper. He agreed to talk to the aftermarket sales team lead and get back to me.

Two weeks later, I was pulled into an informal interview. "So, I hear you're interested in shaking things up around here." I shared my insights on the inefficiency of having to send a representative to the field, then wait for them to come back, tag up with an engineer, and then start the solution brainstorming. My unique experience of starting in the Tooling group, working multiple assembly lines, and having years of experience in design made me a unicorn—not to mention my infectious attitude and rapport with people. A few weeks after that, I was scheduled to move groups and become the department's first-ever Customer Account Manager. I would be managing over 100 international customer accounts in Africa, the Middle East, Southwest Asia, and India.

My first big meeting with a customer was with Ethiopian Airlines. I was so nervous and excited. The crazy part was having to account for the nine-hour difference. I remember having to wake up at 4:00 a.m. to make it to the office for our 5:00 a.m. call. Their engineering and maintenance crews were incredible. They had such great energy, and I was able to build personal friendships with most of the people I worked with. It was truly an honor to serve all my customers, and I really got into the swing of things once I started traveling. My first trip was to Ethiopia to walk their aircraft and evaluate their fleet for many of the issues they were facing.

In March of 2016, I learned I would be traveling to Addis Ababa, Ethiopia to visit Ethiopian Airlines and Nairobi, Kenya to visit Kenya Airways. It was my first time out of the country, and I didn't even have a passport. The company expedited my passport, and I got it in no time. Thankfully, my boss was able to come with me to show me the ropes for my first two trips.

My visit to Africa left me speechless and changed my life forever. It's hard to describe the experience, but the people and culture are so rich and deep. One month later, in April 2016, I was off to Dubai, United Arab Emirates, to visit Emirates Airlines and Abu Dhabi to visit Etihad Airways. Being in the Middle East expanded my understanding of the world. The level of wealth and class of people in the UAE is just astounding. I had the great opportunity to meet some of my international coworkers that I collaborated with in supporting our customers, and they remain some of my dearest friends to this day. I visited the Burj Khalifa, the tallest building in the world, had Spanish wine with a Lebanese dinner, and sang karaoke with Emirates flight attendants.

Two weeks after returning from Dubai, I was asked to support an emergency customer support project and flew to Southern France—Toulouse, France. One month later, in May 2016, I was heading back to Addis Ababa, Ethiopia, for two weeks straight. This was my first solo trip. I was supporting a major supplier meeting and even got to celebrate Ethiopian Airlines' seventy-year anniversary. They had a huge party, introduced me to raw beef (a delicacy), and even had live performers. It was so much fun dancing and celebrating. During my two-week stay in Addis, I got the chance to visit the National Museum of Ethiopia, where the Lucy skeleton is preserved.

Four months later, I was sent to Tashkent, Uzbekistan, to visit Uzbekistan Airways and Ashgabat, Turkmenistan, to support Turkmenistan Airlines. While in Uzbekistan, I learned the majority of the Uzbek people spoke at least three languages or more. A fun memory was meeting a hotel concierge who agreed to be my tour guide. She took me shopping at the local market, and she looked like she could be related to me.

The second leg of this trip took me to Turkmenistan. It was so wildly cold the day they took me to their aircraft fleet that my lead point of contact pulled out one of his extra coats for me to wear. During my return flight home, I had a twelve-hour layover in Florence, Italy, so I took full advantage of the adventure and caught up on sleep during my fifteen-hour flight back. With all this travel, I racked up 60k miles in a seven-month timeframe.

I lived this dream job for eleven months, and then (what felt like overnight) I was let go in January 2017. I could write an entire book on the devastation and mental struggles that came with getting laid off, but I'll fast forward to how this was the perfect timing for my next move. At twenty-seven years old, I was both proud and frustrated at the surreal progress of my career. I thought, *I'm only twenty-seven! Where do I go from here?* I did know that I did not like the idea of someone else having a say in my career, my future, or my ability to be employed.

The benefit of getting laid off is that the company usually provides a severance package. This allowed me to pay for my living expenses while I figured out what to do next. While I was hoping to find another full-time engineering position, it was nice to not be on the hamster wheel of hustle for once. I was able to work out, take time to learn how to become a better cook, and tap into hobbies.

While on a personal trip for my birthday with my future husband, I was on a train going from Barcelona to Valencia when I began journaling. It was such a surreal moment to be sitting there, achieving a bucket list item—all without having a job. Thankfully, we had paid for the majority of the trip ahead of time (Airbnbs, hotels, and flights), but this was crazy! I started to challenge myself to dream. If I didn't want to be stuck in the hustle and bustle of corporate America, what would I do? The idea of owning my own international consulting firm became a clear vision. And thus, in the following six months, I would start BCM Engineered Solutions.

As soon as I started getting my grounding on what it meant to start a business, I submitted paperwork with the state of Texas for a "Doing Business As" (June 2017). I was approached by a recruiter on LinkedIn about a position for a Fortune 150 automotive company. "Your skills perfectly align with our Advanced Manufacturing Engineer role we have here in North Texas. If this is something you'd be interested in exploring, we could really see you being successful here." And she was right on the money. I figured at this point that I could treat this manufacturing company as my test run. I could track all my work, savings, and projects as a case study for what I could do for future clients. Not only was it cool to now be in the automotive manufacturing industry, but I oversaw and was responsible for multiple assembly lines. I sat in the design engineering cubicles, but got to wear the plant uniform of steel-toed boots, Dickies, and a polo or button-up shirt. This was the dream. I was able to oversee the launch of new products with brand new, never before-seen technologies and work on various teams for improvements. This role was the perfect culmination of all my previous experiences, and I was having fun.

Regardless of the job title or position I had served in, the reality about engineering is that it is problem-solving at its core. As an engineering student in college, the topics we learn—calculus III, thermodynamics, dynamics, and mechanical design—are all about finding a solution. For example, calculating the time it will take for the coffee in a cup to come to room temperature. This requires critical thinking and deductive reasoning. Remember those math problems where Sally had five apples, Jose had seven oranges, and David had two apples? We had to learn to negate the information that was not relevant to the problem we were solving.

Learning to explain the world around you is pretty powerful. Some of my favorite classes were geometry and trigonometry. These basic equations could explain how things make sense in the three-dimensional world in two dimensions. You can imagine how much I nerded out once I got to physics and could calculate gravity, force, and friction.

In the world of thermodynamics, or energy conversion, there are tons of graphs, charts, and diagrams. Based on previous findings, research, and laws defined, they all serve as potential tools. Similarly, to build a house, sometimes you need a hammer and nails, other times you'll need pliers, and at other times you'll need a staple gun and measuring tape. Figuring out which tool to use and when is part of the fun. It can be overwhelming, but training your mind to slow down, process, and reason, and then begin taking action, is incredibly empowering. Often, I remind myself about all I've been able to learn. Everything is figure-out-able. In fundamental science, you have the scientific method. You have an idea, an assumption, or a hypothesis. You test your hypothesis. Right or wrong, you try again, tweak some things, list out your findings, and repeat the cycle until you either find your solution or use the findings to inform your decision to make changes to come to a different solution or experiment altogether.

In the world of engineering, we're essentially trained learners. We don't have all the knowledge in the world—we just have a skillset of knowing how to go look. Is this information relevant? Does this chart help us apply more dimensions or structure to the information we're looking for? Failing is a part of the process. You must jump in and start applying things to see if they work. Trial and error are the best teachers.

My expertise comes from my love and compassion for my fellow humans. I have witnessed firsthand the limitlessness of connection, community, and people in synergy and teamwork. Humans are extraordinary! After spending years building processes, step-by-step work instructions on how to complete a task, creating protocols, sharing best practices and lessons learned, and establishing risk mitigation plans (plans for if or when things go wrong), I've learned the more people involved, the better.

Humans are complicated. No two humans are the same. Learning how to navigate, motivate, and get the best of people is not simple. However, utilizing the entire realm of science—the whole toolbox—to build both structure and process is not only amazing but extremely powerful. Historically, sciences have been broken up into "soft" sciences and "hard" sciences. Hard sciences use math explicitly; they have more control over the variables and conclusions. They include physics, chemistry, and astronomy. Soft sciences use the process of collecting empirical data and then using the best methods possible to analyze the information. The results are more difficult to predict.

As both a musician and an engineer, I don't believe in just using the creative "right side" of your brain or just the logical and analytical "left side." Why not use both? Thus, over the past ten years, I have built The People Engineer® as both a brand and an effort to tie the two worlds together. It involves applying engineering methods to everyday people's problems.

In today's world, everyone is focused on automation, efficiencies, Industry 4.0, connectivity, and increasing productivity. And as I mentioned earlier, the most untapped resource is and always will be people. We're truly amazing, complex creations, and the more that we can harness that power and better integrate with the world of technology, systems, and operations, the better off we'll be. Despite all the hardships, I was able to push through; having a true passion for growth, understanding, and curiosity, which saved me. It not only makes me a great engineer but also a better human, neighbor, friend, and consultant.

My favorite memories from any engineering role are the people I was able to serve and how I was able to make their lives easier. It's truly an honor to serve, and as I continue to build out my own consulting practice to be that international consulting firm dedicated to establishing stronger processes and higher performances, it will always be grounded and centered around harmonizing humans. Serving, acknowledging, and empathizing with fellow people is where I start.

Through various leadership roles, I have built over twelve leadership programs, mentored and coached fifty early-career engineers, and led multiple change management initiatives at local, regional, national, and international levels. I've had the great honor of speaking to tens of thousands worldwide and have given hundreds of international workshops, keynotes, and presentations. I've been blessed enough to be featured on national and international television in shows like *Engineering Catastrophes [Season 4]* on the Science Channel, Buzzfeed, and Fox's Domino Masters. I hope to continue sharing my passion for the world of engineering and problem-solving and letting others know what's possible.

So now, dear reader, I challenge you to dream big and reverse engineer the steps you'll need to get there. The world is limitless, and life has endless opportunities. We only know about the one shot we get. The advice my mom always gave me growing up was "everything in life is a choice." You choose. You choose where to focus your energy. You choose whether to see the good, the bad, the ugly, or the lesson. At the end of the day, our experience in life boils down to two options: love or fear. I hope you seek love. I pray you find it, experience it, and share it every single day. Don't be afraid. Push through the fear and just see it as a simple trial-and-error exercise. You're simply testing your hypothesis, remember? I'd love to hear your story of how you leapt and tried. Reach out to me and let me know how it went. Sending all the love, positive energy, and well wishes your way. Ok, bueno, bye!

Connect with Brianne:

GETTING FROM ONE PLACE TO ANOTHER

Coauthored by Vanessa Eslava:
Civil Engineer at TYLin

My name is Vanessa. I was unaware that the profession of engineering was an option for me until a family friend suggested it when I was in my senior year of high school. Although I enjoyed subjects such as math and science in school, it was never recommended for me to become an engineer. After graduating from high school, I took an "Intro to Engineering" course at my local community college, which solidified my interest in getting a degree into engineering.

Through the "Intro to Engineering" course, I discovered the positive impact that civil engineers have on helping build society. Civil engineering is the oldest engineering profession that surrounds us every day. I wanted to choose a career where I could use my interest in math, help care for people, and help my family financially. Now as a civil engineer, I have the ability to work on projects that could improve transit safety for drivers, passengers, and pedestrians.

As a civil engineer focused on transportation, I work on projects that improve the safety and design of pedestrian, bike, and railroad crossings. I spend much of my typical workday as a civil engineer drafting in AutoCAD, a computer aided drafting program, to design my projects. My latest project is focused on cycle tracks that protect cyclists with buffer medians that separate the bikeways from adjacent motor lanes, minimizing rates. I have also designed a pedestrian undercrossing located at an existing track across an elementary school. With the completion of this project, the residents are now able to cross under the railroad tracks safely. The thing I love the most about being a civil engineer, is knowing that the projects I work on help improve the public's ability to get from one place to another safely.

Since becoming an engineer, I realized how I didn't have many mentors who came from a similar background as myself. I began using my social media platform (@itsvnessa) to inspire young girls to see that they can be engineers too, and connect with a community of other women in engineering.

One of the main things that drives me is giving back to the community through K-12 STEM outreach. During the pandemic, virtual outreach has allowed me to connect with students all over the world, including attending a video call with civil engineering students in the Philippines. Although we are thousands of miles apart, it has been great to connect and give advice to students about my experience as a civil engineer.

As I continue in my career path as a civil engineer, I want to be a mentor for the future generation to help build and improve the infrastructure in our society.

Connect with Vanessa:

A PLASTCHICK

Coauthored by Lynzie Nebel:
Plastics Engineer at Cytiva

'm PlastChick, Lynzie Nebel. I'm a plastics engineer from Erie, Pennsylvania. I have three (soon to be four) children with my husband, Daniel. I come from a smallish town outside of Buffalo, N.Y., and from a relatively large, close extended family. I completed my plastics engineering technology degree from Penn State Behrend in 2008 and have been an enthusiastic member of the industry ever since.

Engineering was not the path I thought I would take growing up. I didn't know any engineers, and honestly, I didn't know what an engineer did. Through a series of fateful accidents and misunderstandings, I took a course on manufacturing in high school. I thought the course was required, but that was due to the confusion of communication between myself and a new guidance counselor. However, I ended up loving the class. I thought it was fascinating, almost like a well-kept secret. At the end of the class, I saved my folder, but I figured that was the end of that chapter.

Fast forward to that time in a high school senior's career when they're expected to plan out their entire life. I had always figured I enjoyed history and was a fairly good violist, so my thoughts were torn between the two, leaning toward the viola. A few weeks before I was going to start applying to schools, I was warming up with the rest of the cheerleading squad. I ended up falling from the top of a formation and hitting my wrist. Hard. That injury ended up sidelining me for a few weeks and made playing an instrument for more than a few hours pretty painful. (It still hurts to this day if I try to invest too much time in my viola.) I had to rethink my plans. I took a "What college major should you have?" quiz out of sheer desperation. To my surprise, it spit out plastic engineering as an option. I had literally never heard of that or even considered it for a career.

I started my college research and found there was a school close by that offered a plastics engineering accreditation. This was only one of five schools at the time to offer it. After convincing my parents that I needed to go out of state for my education, they agreed to take me on a tour. One look at that lab and the hum of production, and I was sold. I will always be grateful for the divine intervention that pushed me into a completely different life.

I am currently an upstream product quote engineer at Cytiva. This was a leap for me since the company is in biopharma. The past fourteen years in the industry had been spent in injection molding. I have been involved in modifying products, building molds, and bringing products into production for everything from micro and macro medical devices to dog toys, consumer goods, automotive parts, radiator end tanks, and many other products in between. I have a lot of experience across the process stream, and this felt like the next way to expand my knowledge.

Outwardly, my job is pretty straightforward. I quote any customizations to the hardware for bioreactors and mixers that are used in the upstream part of the cell growth process. The actualities of that mean communicating with commercial departments (and their customers) on what their requested changes mean for their processes, costs, and factories, and then working with a team of subject-matter experts (SMEs) to figure out the materials, time, costs, and processes that will be needed to complete the customer's request. The beauty of a job like this is that you learn about the basic functionality of the equipment and then what it is truly capable of. It's the perfect position to be in when you're learning a new job.

For more than a decade that I've been in the industry, I've been able to achieve some things that the 2004 me would have never expected. In 2018, I joined the executive board of the Society of Plastic Engineers (SPE). I had been an active member of the society since my college years, but always at the local and division levels, and never at the national level. The position was an appointment for one year, and I had to prove my worth with my accomplishments during that year when I had to officially run and be voted in by the councilors of the society. I was able to secure my position, and when I ran again in 2022, I was reelected.

The Executive Board is responsible for the strategic direction the society will take each year, and I am very lucky to be one of the people who gets to be in the room where it happens. I am also the cohost of a podcast called *PlastChicks*. My cohost, Mercedes Landazauri, and I wanted to create an easy and accessible way to introduce people outside of our industry to the incredibly varied career paths in plastics. We highlight awesome people in our own industry, and sometimes we can get caught up in our own corner of the plastics world. We're in our fourth season right now, and we're incredibly proud of the work that we've been doing. That hard work was highlighted with a win from Association Trends for Best Podcast of 2022 and their overall Pinnacle Award.

Part of my work with SPE involves the SPE Foundation, which has recently partnered with the Girl Scouts of Northeast Texas to create a patch. The first patch is called "Color Your World" and shows young girls the excitement of working with color in the plastics industry. There have been big fundraising efforts so we can create a second patch of sustainability. The plan is to continue to partner with other Girl Scout locations so we can expand the program.

Through PlastChicks, Mercedes and I highlight both men and women, but I think it's also important to point out that we have been fortunate to find so many women as well as men who have really amazing journeys in the plastics industry. With SPE, I am also part of the Diversity Equity and Inclusion Advisory Board (DEI), which is working to make sure our industry is doing all that it can to ensure we are encouraging and supporting women (and other minorities) in engineering. SPE has also created a mentorship program that I am co-heading with another incredible woman in the industry.

Before I decided to go to school for engineering, I think I thought of engineering as a much more straightforward field: a few calculations here, punch something into a calculator, build something, repeat. What I've found in the plastics industry is that the opportunities can range from the very classic, calculations-driven engineering job, to the mad scientist in a lab, to the incredibly creative and design-centric, and even to the hands-on "let's fix this" types. There is no limit to where your curiosities can go in the plastics field, which means we need more people who showcase different skills and have different passions. I also believe that I had never considered an engineering path for myself because of the old "you have to be good at math and science" adage that gets tossed around. I don't disagree that a good base level of knowledge in those fields is necessary; however, we never define what "good at" means, and I think for a lot of students, it can deter them from considering STEM fields if they're not the best at one or both of those subjects. To me, engineering is about figuring things out and having the tenacity to not give up. If that gets coupled with curiosity and passion, it won't be long before a student can become successful in the engineering field without being the best math and science person.

There are a lot of challenges that being in STEM and being a woman in STEM can bring. I think almost every woman I've ever spoken to has had some form of imposter syndrome at some point in her career. For me, I think I underestimated just how far imposter syndrome can sink into all aspects of my time in the industry. In college, I often found myself thinking that I understood less or had somehow been misunderstanding the information being taught more than my male classmates.

Looking back, the reality was that I was perfectly in line with the rest of the group, but I just didn't see myself that way. It led to me not answering questions in class and sometimes not stepping up for tasks I probably should have, because I was concerned I would look foolish. Part of overcoming imposter syndrome is embracing the "what if I look foolish" narrative. As I've progressed in my career, the infamous syndrome has also kept me from applying for jobs I was interested in. It's kept me from offering to lead the projects I have a passion for, and it's kept me from strongly backing the ideas I believe in. With time (and some small wins overlooking foolishness), I've been able to recognize these situations where I'm holding back, and start to recognize I have value as a contributor.

The plastics industry is a field that has an incredible need for women's viewpoints. Plastic product design, functionality, and use are all important to consider from both men's and women's perspectives. As a member of this industry, there is a responsibility I place on myself to make sure we are letting women know their voices need to be heard in equal numbers. We make strides every day, but the continued success of programs through SPE and personal commitment from our male colleagues can help bring us closer to that goal.

Connect with Lynzie:

WHATEVER YOU DO DON'T STOP AND DON'T GIVE UP!

Coauthored by Michelle Vargas:
Associate Director, Enterprise Supplier Quality
at Collins Aerospace

My name is Michelle Davis Vargas, and I am from Managua, Nicaragua. I grew up in my home country with my dad (Franklin), mom (Ligia), and my two siblings, William and Franklin. I am the eldest of my siblings. Growing up, I lived happily with my family; we did everything together, and I lived a normal life.

I grew up in Nicaragua, and my dad arranged for my brothers and me to attend La Salle Elementary School. This was a prestigious school that he had attended, and he wanted us to get the best education, just as he had. La Salle is the school that ignited my passion for math and science. My classmates at La Salle were predominantly boys because it used to be an all-boys school. About forty-two boys and eight girls were in each class, and the boys did not intimidate me. At this school, I became extremely good at solving math problems. My fifth-grade teacher held weekly competitions among the students, sending two students to the board, and timing us to see who would be the first to complete the math problem correctly. I remember always being sent to the chalkboard with the brightest boys. Some days I would win the problem-solving competition, and some days I would not. It didn't matter to me since it was a friendly competition that didn't affect our grades. I eagerly awaited these competitions because of my passion for math.

When I was eleven years old, my life changed drastically in Nicaragua. In 1979, the communist government under Daniel Ortega took over. Living under a communist regime affected not only the way we lived but also our education. Food and most toiletries were rationed. Each family was given a card to purchase their essentials, such as sugar, rice, beans, soap, toilet paper, and toothpaste; the records of what was provided were marked on a card. We would run out of the basics around the second week of each month, and we then had to rely on knowing business owners or friends to get the minimum of what we needed. Whenever we ran out of toilet paper, we would go to places such as restaurants and make small folds of toilet paper to bring home.

School life also began to change under the communist regime. I remember that for my history classes, we would have to memorize the names of all the communist leaders, stand up in class, and chant communist slogans. Most of the slogans were against the Americans. One of the chants I remember clearly since I was forced to repeat it regularly was, "*De Aqui o alla, el Yankee morira*," which translates to "From here to there, the Yankee will die." At the time, the U.S. President was Ronald Reagan. The communist government would hang piñatas filled with candies in public parks in the shape of President Reagan. This showed that hitting or hatred of the U.S. president would be rewarded with sweets after hitting the piñata. My family did not agree with communist beliefs or morals happening in our beautiful Nicaragua, and we never engaged in any activity against the U.S. My dad would tell us to remain quiet about our disapproval of the government and to always be careful, noting, "The walls can hear us." Despite living under this regime, we continued to live with and put up with everything happening in our surroundings.

One day, the government mandated that professionals be sent to the mountains to cut coffee and cotton for the government without any pay and under terrible environmental and living conditions. While this was happening, the government ordered that any children from fifth grade to sixth grade would have to go to the mountains to cut coffee and cotton for the government, and those who did not obey would not be able to pass to the next grade in school. My dad was a bank manager and was suddenly told that he was going to be sent to the mountains. When my dad returned from several days in the mountains, he looked at me and said that the government would not send me to the mountains. Many students who went to the mountains returned with diseases and other stories that I do not wish to share. With this mandate in place, in 1985, my parents made the hard decision to send me away by myself so that I could have a brighter future in America and wouldn't have to put up with the communist government anymore.

During that period, I was not the only one trying to leave Nicaragua; most middle-class people also wanted to leave. People were fleeing by the thousands. With everyone wanting to leave, coming to America took work. First, I had to get a tourist visa at the U.S. Embassy. One month before going to the U.S. Embassy for an interview, my mom began to prepare me on how I would answer any questions the U.S. immigration officer might ask me. She would say, "If they ask you this, you respond like this," and we practiced several scenarios. My mom also helped me build confidence for the interview each time we practiced. The day I went to the U.S. embassy, there were long lines that would surround the building for blocks. When it was my turn for the interview, I was shocked when the officer told my mom to go and sit down; he only wanted to speak to me. I was so frightened; however, I could answer all the questions precisely as needed, just as my mom had prepared me. That day, out of the many people at the embassy, only two got a U.S. tourist visa valid for ten years: one other person and me. I felt like I had won the lottery that day!

On February 17, 1985, one day before my twelfth birthday, I left my home country and said goodbye to my family and friends to make my journey to America. I traveled alone. I left Nicaragua with one suitcase in my hand. The suitcase had very few of my belongings, but it was filled with memories of my parents, my dreams, and goals for a better life and education. I arrived in Arizona to be reunited with my Grandma Sara, who had recently arrived from Nicaragua and escaped the same communist regime.

In Arizona, I attended Adams Elementary School and began class at the end of February 1985. I was placed in what remained of the sixth grade. Keep in mind that I had just completed fifth grade in Nicaragua since the school year ended there in December 1984. I knew very little English but still worked hard in school and faced even more challenges to learn to speak English. I needed to memorize unfamiliar words and subjects whose meaning I had no idea of in my native Spanish. I always kept the Spanish-English dictionary with me. Even though I did not understand most of my classes in English, one subject that did not need any translation and was super easy for me, was math. I excelled in math in the sixth grade despite not understanding the language. I completed the sixth grade and went to the seventh grade at Powell Junior High School. When I started Powell Junior High School, to my surprise, students with the same math level as me in the sixth grade were placed in pre-algebra in the seventh grade.

On the other hand, I was placed in the lowest math class possible, general math, maybe because I spoke Spanish and the school thought I couldn't attend an advanced class. This class was a piece of cake for me; I was so bored! One day in this class, my math teacher, Mr. Gilbranson, announced that he would have a math contest. The contest was that whoever got 100 percent scores on their math tests throughout the year would receive a $100 U.S. Savings Bond. That contest got me excited, and I was up for the challenge!!! My goal in this easy math class was to aim for perfection and get 100 percent on every test so that I could win. I figured that getting 95 percent, 96 percent, 97 percent, 98 percent, or 99 percent on my tests would put me behind in the contest. That year, I had nineteen 100 percent test scores. I was the Math Student of the Year out of all the general math classes, and I won the $100 U.S. Saving Bonds contest! My math teacher, Mr. Gilbranson, even gave me a plaque with my name engraved on it for being the Outstanding Math Student.

I continued to get outstanding grades in math and science, which made me realize at a very young age that I should pursue a career involving math. My idea of becoming an engineer became more concrete when I met my computer teacher, Mr. Shuster. Mr. Shuster saw my hunger for education and believed in my full potential. He knew my hardship story of leaving Nicaragua and my family to get a better education. He began to sign me up for minority scholarships to attend as many STEM (Science, Technology, Engineering, and Math) camps as possible at the various universities in Arizona. With his letters of recommendation, I was awarded scholarships every summer in Junior High to attend STEM camps at NAU, UOA, and ASU. These camps were in math, computers, astronomy, and engineering. After attending all those camps, I knew I would be an engineer because I loved all the math and science problems engineers solve to create products, buildings, systems, etc. I narrowed my engineering choice and chose industrial engineering. Mr. Shuster was a catalyst for choosing engineering. He believed in me, which gave me confidence to believe that I could achieve my dream of becoming an engineer. If I studied engineering, I would fulfill my parents' dreams, in addition to my dream, to help the less fortunate by choosing a career that paid well enough.

During this time, my cousin Alfonso had an epiphany, and explained that if I hoped to get accepted to an engineering school, I would have to take more advanced math classes than the easy ones I was taking. He stated that the general math class would not prepare me in school to be admitted into any engineering university. He recommended that I take math classes in the summer to reach AP calculus by the time I graduate from high school. Despite me being an outstanding math student every year in the easier classes, he insisted that that would not help me get into the engineering program at a university. Once he said that, I signed up to take math classes every summer.

I was very poor at the time, and we did not have a car or anyone to drive me to school, so every summer, I rode six miles on my bike in 110+ oF weather to take more advanced math classes. Everyone in the class was taking the class because they had failed the course; I, on the other hand, was taking the class to advance in math and get to engineering school. I aced all the summer math classes I took in junior high and high school.

By the time I attended Westwood High School, I was placed on the track for that grade's highest math level and in the same math class as my gifted friends. In high school, I studied for about five hours daily to get the best grades in all of my classes. Not only that, but I also took on sports, and was part of the girls' tennis and golf teams. I was also very involved in various school clubs. One club, in particular, was Special Buddies, where I would partner with students with Down syndrome or other disabilities, take them to school events, and we would watch them together. We also celebrated various holidays together as a bigger group. I found great joy in helping others. This also gave back to me in significant ways for the career I would study.

At one point in high school, one of my friends approached me and asked if I knew that I was ranked #1 in the school based on my GPA. At the time, I had no idea what ranking was, and he showed me that my ranking in the school was next to my name in a printout of science grades. I then realized what that meant and was very excited, but did not pay much attention to the ranking. I continued to focus on getting good grades and didn't think too hard about the ranking, or a strategy to keep that ranking, for the next three remaining years in high school. I continued to excel in all of my classes while I waited for my parents to eventually make it here, so we could all be reunited and be a complete family again. It had been four years since we had been together, and at the time, there were only phone calls about once every three weeks. My dad would write to me weekly.

My life took a sudden unexpected turn on October 21, 1989, when I was a junior in high school. At the end of 1989, my entire family (mom, dad, and two brothers) was scheduled to get their U.S. green cards. However, our dreams were shattered that day. My beloved dad was returning from a business trip, just one minute from landing in Honduras, when his airplane crashed into a mountain. Historically, this is still considered the worst Central American commercial accident. One hundred and thirty-one out of 146 people died. With my dad gone, the dream of having my family back with me in the States, and obtaining our green cards, was gone.

I took one week off from school to mourn his loss, refocus, and think about what I would do next. My friends were very supportive, and even took the time to speak to my teachers to explain my situation.

Since my dad was gone, I asked myself what his dream would have been for me, and I realized that he definitely would have wanted me to continue to do well in school, and pursue a college degree.

In my junior year, the same year my dad passed, I won the most awards in school. In that year's award ceremony, I won a Bausch & Lomb Outstanding Chemistry Medal, with a full scholarship to the University of Rochester. I was the Junior of the Year, and ASU awarded me a full scholarship. I was also awarded the Outstanding French Student Award in that awards ceremony. I went on to graduate from high school, ranking #3 out of 660 students with a GPA of 4.333. I graduated in the top 1 percent of my class, and because of this, I sat in the front row at the commencement ceremony.

Around this time, I met my husband. We got married when I finished high school and moved to San Jose. My husband, Enrique, has always supported my education. It helped that he was already an engineer and could help explain the more challenging courses to me, such as physics and mechanical engineering. I was accepted as an undeclared major at San Jose State University, since engineering was impacted. I decided in the seventh grade that out of all the engineering fields, I would study Industrial and Systems Engineering (ISE), which I pursued in college.

During my second semester at San Jose State University, I showed up to an industrial engineering class that I wanted to add; the course was packed, and I could not even be wait-listed since engineering was not my major. In that class, Professor Freund announced at the beginning of the lecture that he needed a student to help him establish a mentorship program. As all the students were leaving, I approached Professor Freund and offered to help him with the mentorship program. After talking to him for about an hour about my passions and goals, not only did he add me to his class, but that same day he made sure that I was enrolled in the ISE program. Dr. Freund also made me the chair of the mentor program. I saw the value of a mentorship program. This was the origin of mentorship, which later in life became a big passion of mine. I didn't let any obstacles stop me from getting into engineering school, but instead, always figured out a way to achieve any goal I set my mind to. I have always had it in me to never give up and to always look for opportunities to get to where I want to go.

Dr. Freund was also a catalyst of my engineering dream; without knowing me, he gave me the opportunity, and signed me up for engineering. How cool was that? My engineering classes were filled with mostly guys, and for the most part, about three ladies in each class. From my early La Salle days, I was used to being in classes where most of the students were male, and once again, I didn't let this intimidate me. I focused on getting good grades and ignored any difference between how my male peers treated each other, and how they treated me. I just blocked it out of my mind.

In my first year of college, I joined the IIE (Institute of Industrial Engineers) and became vice president. I also joined the SHPE (Society of Hispanic Engineers) and SWE (Society of Women Engineers), and remained active in all three clubs. I can't say that engineering classes were easy because they were not. However, since I had a husband who was an engineer, he would explain my homework to me with real-life examples until I could understand it. That helped a great deal because I had a tutor and mentor at home.

In my third year of college, we decided to have a baby; and my daughter Saralynn was born in the summer of 1994. While pregnant in school, I would nap between classes on the IIE club's couch. The males in my classes wouldn't understand why I had to sleep so much, and would even ask me if I had a home; I would again ignore the comments. Since my daughter was born in the summer, I did not stop attending class the semester before or after her birth. Despite having a baby, I excelled in school, mostly getting As and Bs. It took me five years to obtain my degree, and I eventually graduated *cum laude*.

My first job out of college was as a quality engineer for LifeScan, a Johnson & Johnson company, that manufactured glucose monitoring devices to help monitor glucose for people with diabetes. I landed in the field of quality by accident. When I was about to graduate from college, I asked a student friend if he knew anyone hiring engineers. He gave me the business card of Eric Veit, who was the quality engineering manager for LifeScan at the time.

My friend owned a gas station. One day he had his statistics book opened, and Eric, who was filling his car with gas, started talking to him. Eric has a masters in statistics, and this grabbed his attention. He chatted with my friend and gave him his business card. My friend gave me Eric's business card, and I was under the impression that they were close friends, rather than two strangers that met and had a conversation. Therefore, I called Eric confidently, gave him my background, and showed my interest in working at LifeScan. Eric allowed me to interview for a quality engineering position, and I got the job! The actions that I took, once again, show that by having confidence and using connections, I could land a job at a very prestigious biotech company.

From my first job, I realized that I always wanted to work in the biotech field. Creating products to help others and the community is something I have done since I was a little girl. I have worked for over twenty-five years in biotech (medical devices and pharmaceuticals) and, more specifically, in the field of quality assurance (for most of my career as a quality engineer). I worked my way "up the ladder" and excelled in the individual contributor path as a quality engineer. I moved from quality engineer and was promoted to the next highest level, senior quality engineer. I was then promoted again to one of the highest levels, staff quality engineer. I pursued these leadership roles five years ago because I like working with people and leading large technical teams.

My first opportunity in a supervisory role and leading a group of employees was at Scantibodies Laboratory Inc.®, where I was promoted from quality engineer to quality supervisor. Later I applied to work at Illumina, the #1 biotech company in San Diego, and was hired as a staff quality engineer. During my years at Illumina, I worked hard and interviewed when an opportunity opened for the manufacturing operation's quality manager position, and I got the job.

Fast forward to today, and I continue to be a lifelong learner and continuously educate myself, obtaining certifications in quality assurance. I even returned to school to get my Masters in Regulatory Affairs from San Diego State University (SDSU) in honor of my father, who had a masters degree. Still, I felt that I had to continue and carry on with my profession where he left off with his masters.

Continuing to educate myself even after college graduation and growing in my field gave me the tools and confidence necessary to apply for an upper leadership position as an associate director of quality systems at Pacira Biosciences. Working for medical device or pharmaceutical companies requires that the company and its employees follow regulations (or laws) by the FDA® (Food and Drug Administration), which is under the executive branch of Congress. The FDA's primary role is to protect the public's health by ensuring that drugs and medical devices meet quality standards and are safe and effective. Pharmaceutical manufacturers follow the FDA's Quality Systems Regulations titled 21 CFR 210 and 211. Medical devices follow the FDA's Quality Systems Regulation 21 CFR 820. There are many other laws and regulations that the companies that I have worked with have had to follow to ensure that the products that we made functioned as intended. These products help society in the treatment, diagnosis, or cure of illnesses.

As an associate director of quality systems, I managed three teams: the training team, the document control team, and the electronic quality systems team. There are FDA regulations associated with each of the three groups. My team's role was to ensure that everyone in the company, at all of our locations, met the requirements for training (this included procedures listing the employees' duties, to having the right background to perform their jobs which require having proper documentation), the requirements for documenting control (this included how the company will electronically store the procedures to perform work functions, procedures to meet the various quality system requirements, approving the procedures by qualified employees, having password protection and control so the procedures do not get changed without the proper control), and the requirements for the quality systems team (including owning, administering, and testing electronic quality systems to ensure the data is safe, cannot be changed, and is accurate). Following all the laws and regulations allows biotech companies to provide safe and effective products for patients taking or using prescription drugs or medical devices.

My twenty-five years of working in quality assurance have been gratifying. I chose to work for biotech companies where we make a difference in the community by making products that cure, diagnose, or treat diseases.

One of the highlights of my career was implementing a company-wide validation system at LifeScan, a Johnson & Johnson® company. This was recognized within the company and earned Best Practice recognition in a validation magazine named *Institute of Validation Technology* (IVT). Another career highlight included implementing SPC (Statistical Process Control) for the entire company at Biosite, which resulted in my team receiving the Outstanding Achievement Award. At Illumina I implemented the medical device FDA® regulations company-wide, and across all locations, that were moving from primarily creating research and development devices to commercial ones that follow 21 CFR 820.

Lastly, in my role as the associate director of quality systems, this year my team accomplished updating a global electronic documentation system from a 2011 version to the current 2022 version. Many leaders before me had this project, but none could complete it. My team also migrated all the electronic documentation systems (a 2011 version) from a company we acquired, and merged the documentation into the new 2022 version. This required a lot of knowledge, understanding, and technical skills to prove that all these electronic systems met the FDA® requirements for the software lifecycle, and that all the user requirements, design specifications, and validation deliverables all passed. The entire process was documented as mandated by the FDA regulations.

Working in quality assurance has its challenges since this group's primary role is to ensure that all the laws, regulations, and standards are consistently implemented and followed throughout the company, to guarantee that they give quality products to consumers. It is our responsibility to check these systems daily and notify, stop, modify, test, or correct them when they do not meet the requirements set by the FDA® or the company, and to ensure they are documented appropriately with enough data, and approved by the proper departments in the company. The quality assurance role can be very challenging, as some employees question why some rules are implemented. It is our role to explain, convince, and prove why they are necessary and required by the government and the customers.

Another challenge I have experienced throughout my career is being one of the few female engineers in a male-dominated field, in every company and team I have ever worked for. There have been occasions where I have been in meetings, and my ideas have been ignored, and/or a colleague would share my idea and take credit as if it were his, when I hadn't yet been given a chance to present my thoughts. I have started providing my ideas before the person talking ends his conversation so that I could get a chance and they would listen. I have also been in situations where others with the same or less background and experience have been promoted to better positions with higher titles. In those cases, I continuously look for opportunities, whether it is switching to another group within a company or going to another company.

Later in my career, I wanted to do more to promote women in engineering, teach them how to navigate a dominantly male field, and let them learn from my twenty-five years of experience. The opportunity came in 2016, when my daughter started to attend San Diego State

University (SDSU). To thank SDSU for being so welcoming to my daughter, I decided to volunteer my time and join the AMP (Aztec Mentor Program). Because I was working full-time and have a family, I decided to be very selective about whom I chose to mentor. I narrowed my selection down to mentoring female underrepresented engineering students; mainly first-generation college students. What I wish for them is what I told my younger self when I entered the professional world: to always show confidence in front of others despite feeling the opposite way; to always stand firm in their beliefs and question people in leadership positions, yet do so with respect; and lastly, to speak up when they see unfairness with others or themselves, rather than go with the flow and do nothing about it.

I have mentored over twenty students, and all of them have already graduated with an engineering degree. I am passionate about this and acquire new students each semester. This semester I started to mentor three new students from Southwestern College, a two-year college. As with my SDSU mentees, I guide these students to pursue attending a four-year university after getting their associates degree.

As part of my guidance to all my mentees, I provide tips on studying techniques and other areas, such as helping them prepare their resumes, apply for scholarships and internships, build their confidence and self-esteem, and prepare for interviews. I also encourage them to join not only the student chapters of organizations such as SHPE (Society of Hispanic Professional Engineers) and SWE (Society of Women Engineers), but also the professional chapters of these same organizations, and encourage them to take on leadership roles. I have also sponsored them for engineering events and conferences in San Diego and Los Angeles so they can interact and network with professionals in their fields. Many of my mentees have expressed that if they had not had a mentor like me, they would have given up on school or taken any non-engineering job after graduation. When I give these students my time, all I ask is that they pay it forward and do the same for the younger generations.

As a result of all the help that I have provided to these engineering students and for displaying excellence in my field, I have been the recipient of several awards, including the Outstanding (Female) Engineer by SWE, Outstanding Engineering Service by SDCEC (San Diego County Engineering Council), Most Influential Women in Engineering by the *San Diego Business Journal,* and the NAWBO Brava Outstanding Corporate Leader Award. Despite these engineering awards, I want to be best remembered for how many students and professionals I helped to accomplish their dreams and reach their full potential.

My journey to becoming an engineer has been very challenging, with many obstacles to overcome, but that has never stopped me from achieving any dreams or goals. I continue to face new challenges in my professional career, and I have always managed to overcome them with resilience, perseverance, and by never giving up.

I want to continue to share my story so that the future generations of female engineers or those in the STEM fields know that they, too, can accomplish anything they want if they set their minds to it, continue moving forward, and stay strong even when faced with very tough challenges and downfalls.

My motto in life, and how I conclude with my story on becoming and being an engineer is, *"Aim for a goal; if you land on the target, tell yourself that you have done a fantastic job and give yourself a tap on the back! And if you don't, throw again, aiming for the target, and repeat only if needed until you hit that target. Whatever you do, don't stop, and don't give up!"*

"If you can dream it, you can do it."

—*Disney*

Connect with Michelle:

A JOURNEY OF DISCOVERY: FINDING PURPOSE IN INSPIRING OTHERS

Coauthored by Kelly Kloster Hon:
Sr. Manager, Product Development Engineering
at Becton Dickinson (BD)

I am a steminst. I believe that women play a vital role in STEM (Science, Technology, Engineering, and Math) and shaping our future—we have lived far too long in a world designed for and by men without the female perspective. I also believe that when you have a passion for a cause, you can truly make a difference if you invest your time, talents, and money toward it. Now, I was not always a changemaker—I did not know that if I did not start creating opportunities for myself and other women, that change was not going to happen. Let me walk you through my journey and a few of the experiences that have shaped me.

I have always been a bit of a rebel and a bit of a challenge seeker. I tend to view almost everything as a competition; if there is not one, I try to make one. As a child, I had many interests similar to the boys in my class — from sports to cars, and math to LEGOs — and loved to compete against them, showing I was just as capable. Growing up, I attended the German International School, Washington D.C. (GISW), which offered rigorous academics via a challenging German curriculum that focused on critical thinking. I thrived in this environment and naturally excelled at math, where I had a healthy competition with several boys, comparing our grades after each test. Nothing brings me more satisfaction than the pride of accomplishing something difficult.

When I switched to an American school in the fourth grade, math was the one constant in my life. I continued to excel, but also started to explore STEM more on my own. I had experiences like the middle school science fair where I built and tested a hovercraft, and I attended a summer camp at the University of Maryland, where I built, coded, and competed with a LEGO robot. Simultaneously, I was inspired by some strong female characters in media—from Disney's *Motocross*, where a woman came out on top in a "man's sport," to *Alias*, where Jennifer Garner could dodge bullets, speak a myriad of tongues, and save the world, all while juggling the demands of grad school, to *The Italian Job* featuring Charlize Theron, who portrays a confident con woman that not only outdrove the guys in her MINI Cooper®, but also showed how strength can come from our mistakes and vulnerabilities. Throughout this time, I gained self-confidence and developed the mindset of when you believe in yourself, others are more likely to believe in you too.

Finding engineering as a female can be tricky, especially since it is not a typical field that women are encouraged to consider. Furthermore, it was also not something that, growing up, children normally identified as a profession compared to a nurse, teacher, or secretary. In fact, when asked what I wanted to be when I grew up, my answer was a racecar driver. However, I had a physics teacher in high school that noticed my natural skill set and curiosity in her class and encouraged me to consider engineering. She helped me start a "Junior Engineering Technical Society" (JETS) team at my school, which included a one-day competition that applied math and science knowledge in practical, creative ways to solve real-world engineering challenges.

Around this time, I took an aptitude test to identify my inherent strengths, natural inclinations, and how they aligned with professional careers. Thinking I might move into finance or something math related, the test results came back and helped me pivot to engineering. While I would have excelled in finance, my full potential would have been untapped, and I likely would have been bored quickly. Architecture or engineering was in my future, given my spatial reasoning skills and other aptitudes. **Turns out engineering is in my DNA!** My maternal side of the family has a lineage of engineers, from mechanical, to nautical, to aerospace. My mother probably would have been an engineer had she known about it; however, she majored in math and became a computer programmer.

With this information in hand, I eventually applied and was accepted to the #2 mechanical engineering school in the U.S. at the time—the Georgia Institute of Technology. **Why mechanical engineering?** Well, my mom and I read through the descriptions of the various engineering disciplines and quickly narrowed it down to mechanical engineering. I appreciated the diversity of fields one could go into with this degree, plus the idea that if I could not race a car, maybe I could engineer it!

My journey within mechanical engineering exposed me in many ways to the breadth of careers possible with this degree. I started off in the more traditional space of heating, ventilation, and air conditioning (HVAC) with my first internship. This exposed me to the areas of thermodynamics, fluid mechanics, and heat transfer, as well as what it is like to work for the U.S. government. After an educational experience abroad in Germany, where we visited numerous operations plants from porcelain, to medicine, to laundry detergent, to automotive, I wanted to learn more about how things are made. I therefore accepted a manufacturing internship the following summer in a pressure packaging plant, working on optimizing the TRESemme® Hair Spray production line for a privately owned company. Ultimately, this experience directed me into product development, as I was more interested in designing the products that were being made rather than making them. After graduation, I completed my third and final internship with a consumer goods company, designing and testing baby strollers. You can truly do a multitude of things with a background in mechanical engineering.

During my senior year, I completed my favorite course, Senior Design. This is where I was introduced to a design and consulting firm, IDEO, that uses a design-thinking approach to design human-centered products. I was paired up with two industrial designers and another mechanical engineer to work as an interdisciplinary team to design solutions for a problem Delta Airlines flight attendants were facing at the time. Leveraging the mindset that complex problems are best solved collaboratively, we worked together to understand the situation to ideate, innovate, and implement our designs, some of which can be found in the friendly skies today. I loved working with a team to develop solutions to complex problems and could not wait to pursue a career doing this, after creating a strong base of mechanical engineering skills with a master's degree.

During my sophomore year, I had decided to sign up for Georgia Tech's five-year BS/MS program, as it would enable me to complete a graduate degree with only one additional year of schooling, and without requiring the Graduate Record Examination (GRE). I was never great at multiple choice tests, so I was excited that my excellent academic performance enabled me to be automatically accepted. As I neared my graduation, I decided to pursue my master's degree in product development to gain the additional knowledge and specialization I felt I needed to succeed in the real world. I ended up completing the degree at both Georgia Tech and the University of Stuttgart in Germany with a strong foundation in the design process.

Upon my postgraduation, I applied to jobs all around the world for roles that would provide me with an interdisciplinary product design environment. I ended up back in Atlanta for my first full-time role with a firm called Big Bang, a group of design, interaction, and engineering experts creating extraordinary experiences. We were a tight group of ten people partnering with domestic and international companies and an individual inventor to drive maturity in their businesses through engineering solutions, testing, and consumer research. Here I worked in a fast-paced environment on seventeen different projects over nineteen months—from defining unmet needs that drove key strategies, areas of improvement, and new product ideas to designing and testing complex engineering systems. I gained varied experiences, from honing CAD skills in SOLIDWORKS® to conducting user research, testing products in a cadaver lab, managing schedules and leading cross-functional teams, presenting to customers, writing proposals, and visiting potential clients. In short, I experienced it all.

With my first full-time job came many career highlights, like the day I saw a dental product I engineered hanging on a rack at Walmart, or the day I signed the paperwork for one of my first patents—a sternal retractor, which is used to hold the sternum open during open-heart surgery. It also came with its own challenges, like deciding when the right time is to move on and leave a company, or what it is like to question your career choice while being the only female engineer.

I remember someone telling me once that while we graduate a decent percentage of women in STEM, we lose them in the working world within the first five years of their careers. Even after overcoming the hurdles to enter the profession, women leave at much higher rates than men. The challenges and stress that come with being a female in a male-dominant environment, where women feel that our contributions are less valued, where there is a lack of female role models, and where minimal support exists, make it easy for many to leave the profession. All of these increase feelings of not fitting in at work, so we lose an extreme percentage of women to other career paths such as teaching, real estate, and marketing. I too started to feel some of these challenges and stresses in my first job; however, given my competitive nature, I made it a personal goal to make it past that five-year mark (and I made it!).

While I did not have female role models at work, I did have them through the Society of Women Engineers (SWE), whose annual conference quickly became my annual pep rally. I was introduced to SWE during my first week at Georgia Tech, where I attended "Tea with the Dean," a SWE-organized event where all incoming freshman women were invited. It was rare to be in a room surrounded by so many smart, diverse, and excited women who all aspired to be engineers one day. The energy of the event and the inclusive, opportunistic, and fun demeanor of the existing SWE members got me hooked. I wanted in and have been a member of the organization ever since. The annual SWE conference is my time to be inspired by all the incredible women who are pursuing an engineering life as well as to reconnect with my favorite alumni and SWE friends. It just makes everything feel right again.

With a desire to strengthen my engineering foundation, I switched companies and became a design engineer for Newell Brands. I designed and engineered several consumer products, but my main role was to lead a product refresh on DYMO® label makers, bringing the look and function into the 21st Century. With the "product intent" from marketing in hand, I partnered with industrial design, human factors, electrical engineering, and software engineering to hone in on the product requirements and finalize a design. I used Creo®, a computer-aided design software, to design the product, ensuring all requirements and manufacturing complexities were fully incorporated. This included mold flow analysis to finalize gate locations, parting lines, drafts, conducting tolerance stack-ups to ensure component lineup, and executing finite element analysis to confirm snap features, screw locations, and rib designs. You do not learn everything in school, so I have been very fortunate to have had outstanding key mentors throughout my career. They have patiently coached and guided me as I built up my skills while giving me the autonomy to fail and learn from the experience.

I have found that **highly diverse networks have also helped me develop** more complete, creative, and unbiased perspectives and push my existing personal limitations. When you trade information or skills with people whose experiences differ from your own, you provide one another with unique, exceptionally valuable resources. For example, my opposite is Haley. Haley flies by the seat of her pants. She has no prior experience and runs for president; she barely speaks German and signs up to study abroad in Germany; she presents at a conference with mere minutes of preparation. She inspired me and still does to stretch myself, go out on a limb, and try something I feel completely ill prepared for. She taught me how to approach situations differently; to be comfortable while being uncomfortable. And while perfection does not build character, recovery does. I am so fortunate to have had her and others form a new perspective and approach for me.

After my husband completed his doctorate degree, we relocated to San Diego, where we currently reside. With a small personal network in the area, I evaluated where the Georgia Tech alumni were employed and ultimately found a role at Becton Dickinson® (BD), a global medical technology company. I find that I succeed in environments where opportunities are constantly being discovered and challenges are revealed so they can be overcome. I was transparent about this with the hiring manager, who assigned me to the "Quick Wins" team, where that could take place. I led 75+ cross-functional team members (R&D, quality engineering, marketing, medical affairs, regulatory, and operations) from design and planning through manufacturing and quick commercial launch of multiple products. Having focused on the front end of the product development process to date, I was excited to gain experience with the back end, moving from design into the launch phase. Additionally, there were many new aspects to working in the medical space for me, like the design control process, but I took it as just another challenge to be tackled.

I achieve feelings of self-worth by being a successful leader of others and by providing direction to achieve results. I leveraged SWE to grow as a leader, develop my leadership skills, and make the shift from that of an individual contributor to a manager. Let me tell you—that shift is hard, especially for someone who enjoys crossing items off a checklist! Becoming SWE's President for the San Diego Professional Section afforded me opportunities to: 1) build effective strategies; 2) manage conflict; 3) influence others; and 4) foster teamwork. I had the opportunity to define my authentic leadership style and gain experience before it became my "day job" at BD. I had to relinquish control and trust others. I received critical feedback and needed to adapt. I encountered abnormal situations and reached out for coaching. I tried new things and chose to try again, all within the safe space of SWE.

In my journey to become a manager and now senior R&D manager, **I have seen the criticality of sponsors.** It is important to build relationships and cultivate them so that when you are seeking an opportunity, you do not do so alone. Sometimes you are completely unaware of an opportunity that is perfect for you, but people whom you have invested in and whose lives you have impacted will remember you.

Grisel and I met through the work I have done for BD's Women's Initiative Network (WIN), where she was the lead. She saw the work I had done to compile and submit BD's packet for a SWE award five years ago, and then was able to connect with me in person at the awards ceremony. Our relationship has continued to blossom over the years. Almost two years ago, there was an open position as the global STEM lead, and she is the reason I now hold that title.

In my role as WIN+STEM lead, **I desire to create an environment where women feel as though they belong** in STEM. I focus on amplifying the impact women in STEM have had on driving change. For example, I organized a campaign for International Women in Engineering Day, highlighting the stories and achievements of women in STEM. I collaborate annually with BD's law group to encourage all female associates who have received patents to apply for an external recognition award. I honor women in history who have paved the way through social media. Additionally, in this male-dominated discipline, we need male allies to take action in their own lives to create a more equal, less biased place to work and live. To help us all learn and grow together, I launched an award-winning series to educate associates on the importance of allyship, and equip us with tools and actions on how to become better allies in our everyday lives.

Simultaneously, I am **working towards a future with gender parity and equality in engineering** by building a pipeline of STEM women that we can look up to, be inspired by, and aspire to be. One of the biggest challenges we face is a lack of female mentors and role models, especially in leadership positions. When women do not populate those critical roles, we cannot make progress in a meaningful way. I have launched and delivered impactful professional development events such as keynote speeches on topics such as "How to Attract an Advocate," conducted quarterly networking lunches, and organized skill-building presentations. I have supported other women in STEM who are on their own journeys of making a difference for women in STEM, like my friend Wendy, who has released a best-selling book on Amazon on work-life balance. I like to say, I open doors, push women in front of me through them, and pull women behind me along with me because we need more women at the top! I believe that the best leaders exist to lift those around them, and they do this by helping to unlock their potential. I aim to be such a leader because I know it will change our future landscape and the norms of today.

Lastly, I dedicate my personal time, experiences, and money to helping students develop a passion for and pursue STEM. Teaching young girls about STEM and seeing the joy that solving problems brings them is invigorating. Talking to high school and college students about my varied career path and seeing the relief they feel in hearing that it does not need to be fully flushed out or linear is rewarding. I have educated and inspired hundreds of students, from hosting booths at local STEAM fairs and FIRST robotics competitions, to engaging students virtually and in person through tours, hands-on learning events, and speaking engagements. Some of my favorite events to date include organizing a summer camp for students to visit and learn from different local companies, building solar cars with high school students, and being the keynote speaker at a college graduation SWE event. Investing in our future generation brings me great joy, and I feel very fortunate to be able to also fund multiple scholarships with my husband that support STEM students. We hope that they will encourage youth to continue pursuing engineering and help mold them into passionate, empathetic, and generous global leaders.

As you can see, my journey has included a diverse and global upbringing with varied academic and professional experiences. It harnessed a strong network and leveraged personal experiences to find my passion for supporting women in STEM. I hope my story inspires you to go find your purpose, grow as a leader, and give back to your community. Form your story and fill it with experiences that will shape you.

Connect with Kelly:

EXTRAORDINARY ENGINEERS

Coauthored by Dr. J. A. Sanchez:
Quality Engineer II at TÜV SÜD America

There are moments in life when we get to experience things that allow us to walk away changed. We walk away with a new perspective and know we have to do something. That's exactly what happened to me after a STEM Gems summit. I was so incredibly inspired by what I had just heard and been a part of. I felt this stirring inside of me. Shortly after, I was going for a run and had a divine thought that went like this: "Why not capture all these amazing engineering journeys in a book? Surely others will be inspired like I have been." Thus, Extraordinary Engineers was born.

Allow me to introduce myself. My name is Justina Sanchez. I am a wife, a mom to a little girl, and a quality engineer. I was born in San Diego and raised by my amazing and supportive mom. Growing up, I wanted to be an actress, so I did a lot of theater and took a lot of professional classes, like on-camera acting, commercial basics, and drama classes, to name a few. I moved to L.A. right after I graduated from high school with my then-sweetheart, now husband. I pursued that dream in my heart for years, only to find that the longer I did it, the less I wanted it.

When I was twenty-two, I moved back to San Diego and got a "real" job. I started off as an administrative assistant at the company I still work for, TÜV SÜD America. When I first started at the company, I didn't know anything about engineering; I just needed a decent job with benefits; I got that and sooooo much more! Two years after I started, I began to realize there was this whole world of engineering that I never knew existed. I saw that it was fun and exciting, and no two days were alike. I began to get glimpses of what my future could be if I were an engineer, and immediately my eyes and heart would light up! The more I thought about it, the more my heart would leap out of my chest with excitement!

There came a day when I walked into my boss's office and asked him what I had to do to be an engineer. He replied, "Get a degree in electronic engineering." I said, "Okay," and turned and walked out. Immediately, I went to my desk and googled electronic engineering.

ITT-Tech was among the names that popped up. I knew some of my team had gone there, so I called them up and made an appointment to tour the school the following week.

After touring the school and talking with colleagues, I decided that taking the less traditional route of going to a trade school was best for me at that point in my life. I knew I was working for a great company, and if I wanted to move up in my role and responsibilities, I needed that technical knowledge, training, and experience. ITT-Tech was a great school that immersed me in the world of technology and electronic engineering. I literally knew nothing about computers and electronic engineering, and came out knowing the ins and outs of electronic circuitry and components, along with computer programming and much more. In 2010, I graduated with a degree in computer and electronic engineering technology.

Three months after I graduated, I was promoted to the product safety test technician, and then I found out I was pregnant. This was all new for me. At TÜV SÜD, we are in the testing, inspection, and certification industry. Simply put, we test and certify products before they go to market for consumer use. In my new role as a product safety test technician, I was responsible for performing all the electrical safety testing of products before they went to market. We do tests people usually don't think about while using the products every day. Did you know that almost everything you plug in or that is battery-operated must be tested and certified before you can buy it? Things like flat irons, power supplies for your laptop, medical device sterilizers, and soldering irons and spas. Just about EVERYTHING must be checked for safety before people can use them.

We do some really cool tests in the safety lab. To name a few, we do dust and water testing if a device is capable of being used outdoors. In another test we measure the time it takes for the energy to dissipate from the prongs on the plug of a device, and compare that to the allowable limits indicated in the standard. Ultimately, we want to make sure that no one gets shocked if someone unplugs the device from the outlet and touches the prongs. We also have cool machines that simulate something dropping down a flight of stairs, or tilt ramps to make sure tall products like vending machines won't fall over on someone if they are tilted to a certain angle. These are just a few of the safety tests we do on products during the testing phase. In this role, I was also responsible for all of the lab equipment and operations. During this time I was getting my feet wet in the world of safety testing and getting familiar with how to run a fully accredited lab. (This would become a valuable experience years later.) All the while, my belly was growing each day. I embraced all the new changes in that season of life.

During this time, corporate decided to renovate the lab, and I got to be a part of redesigning it. We had to determine where to put what types of electrical drops, which testing rooms and chambers were best placed for efficiency and functionality, and whether we had the power capabilities to operate the test equipment. Many days I found myself wearing hard hats and operating power tools, with my baby growing and my belly getting bigger with every passing week. After a few years of maintaining the lab and getting familiar with all of its processes and procedures, I felt confident that I could run a fully accredited lab. I even mentioned to a coworker during an audit one time: "I want to do what the auditor does. I want to audit labs." Little did I know that all that time I was being prepared for the role that would come my way eight years later.

In 2015, I was approached by another division's manager. He offered me a role as an EMC engineer. I had never really thought about being an EMC engineer, but it was a step up from being a technician. I brought this information back to my boss and our director, who then made a counteroffer for a product safety engineer position. I thought about the options in front of me and opted for the Product Safety Engineer role. I wanted to be loyal to my division and team, and product safety was what I knew. It was comfortable for me. Now I would manage full product evaluations from beginning to end.

In a typical evaluation, you do a construction review, meaning we have determined what standard the product will be evaluated to and made a list of critical components. Critical components are components that can potentially cause a hazard—whether it be an electrical shock like I mentioned above, energy, fire, heat, mechanical, radiation, or chemical hazards. The idea is to make sure those critical components have all their proper ratings and approvals for use. Some examples of critical components are things like power supplies, fans, batteries, wires, transformers, and printed wiring boards and enclosures, to name a few.

Next, we determine all the production facilities that the product will be produced in, and review all of the block diagrams and schematics in this phase as well. We do this to avoid any possible redesigns prior to testing. The product safety engineer will then compile a list of tests the product will undergo. This is based on the type of product being evaluated. Because every product is different, every evaluation is different too.

Next is the testing phase. This is where the fun really happens! If I am doing an evaluation of an outdoor light, I would run a dust test on the light. This represents sand, dirt, or debris being kicked up around the light. After the test, I will open the unit to make sure no dust got inside the enclosure of the light. If it did, I must consider how much and where, while taking into consideration things that could impair the proper functioning of the unit or impact the creepage and clearances measured.

On another test sample of the light, I would also run a water test because, likely, at some point in its final installation, it would be rained on or exposed to water from a sprinkler system or water hose. After the water test, I again open the unit and make sure no water got inside, or if it did, where and how much. I have to determine if the water could cause a short or pose a hazard to the proper functioning of the light. All of this and more are considered in the compliance criteria.

Another aspect of an evaluation is the factory inspection. Remember I mentioned that during the first phase we determined the production facilities? Well, now it is time to go inspect them. This part of the job is really fun because you get to travel! I get to go onsite to production facilities and make sure the manufacturer is producing what we have approved for use. Some production facilities are amazing! Many facilities are huge, and some have robots that help with the production, while others have people. It is amazing to see the inside workings of how things are made. This is one of my favorite parts of the job.

The final phase of an evaluation that the product safety engineer is responsible for is the reporting, review, and certification (if applicable). If all the open issues have been appropriately addressed, I can write a passing report, which includes all the data from each phase. If there are still open issues, I write a negative technical report outlining all of the "failures" for the manufacturer to address.

If certification was applicable and we have a passing report, now is the time to issue the certificate to give to the manufacturer. This is what most manufacturers need to sell to various markets all over the world. One of the coolest parts about being a product safety engineer is that after you do a product evaluation, you can later see that product out in the world being used. I have often been at the mall or the airport and have seen products I tested and evaluated. I see a lot of X-ray scanners and baggage scanners, and I think, *How cool is it that I was a part of bringing that to market, and here it is being used by thousands of people every day?*

I do have to tell you it is not always easy; in fact, many days I was discouraged because it was hard for me. I felt like an imposter at times (imposter syndrome became a panel discussion at an America's Women's Network event) because it was so challenging. I guess I thought I was supposed to know it all, but later realized that was a ridiculous standard I was holding myself to. Trying to make sense of all the technical requirements is not something that came easily to me.

There were when days I wanted to throw in the towel; I am so glad I didn't, but rather just pushed forward. I asked my colleagues a lot of questions, I took many technical trainings, and I also had a great boss who never made me feel dumb for asking questions. He was, and still is, very approachable; his philosophy is to teach me how to catch the fish rather than just give me the fish, so next time I would know how to do it.

While doing all of this in my professional life, I started searching for something in my personal life—only I didn't know what. I had a beautiful family and a thriving career, but I felt this void that I just couldn't fill. In my search, I started going back to church, and during that time, God got ahold of my heart. I was so hungry to learn and grow in the knowledge of who God is. I spent several years taking pastoral support classes, Christian leadership and development classes, attending the Global School of the Supernatural, and so much more. All the while, I saw myself growing into who God was calling me to be. I learned that with Him, nothing is impossible! We really can do anything we put our minds to. I discovered that the desire in my heart and mind was just a part of the very purpose for which I was created. I am happy to share that in September 2021, I received my Doctor of Divinity from HSBN International School of Ministry! This happened about six years into my role as a product safety engineer.

About the same time, the opportunity to be a quality engineer arose without me even looking for it. It just fell into my lap. The way it happened was like nothing I had ever experienced, and I knew if I didn't go for it, I would have regrets. So here I am as a Quality Engineer II. Now I get to go to all our labs in North America and audit them to make sure they follow our accreditor requirements. It's crazy to think that for years I was testing in the lab and being audited, and now I am the auditor. It's great, and I am loving it! It is very fulfilling and rewarding to know I help to keep the world a safer place, one product at a time. I believe it is all a part of me pursuing the purpose for which I was created!

My heart has always been to give away what I get in every area of my life; whether it be of my time, talent, or treasure. In 2018, I was selected as one of four women globally in our company to be mentored by a CEO. This experience was life-changing for me. My mentor saw something in me that I didn't see in myself. I was so grateful for the opportunity, and I knew I had to give it back. Not long after my experience as a mentee, I attended the Society of Women Engineers' (SWE) open house, where I sat at the same table as a young girl named Madalyn, and her mom, Tracy. Madalyn and I instantly bonded over her Robotics Team, Spyder, known for being a safety team, and because of my role as a product safety engineer. We connected that night, and I eventually became her mentor. She is an amazing young woman who inspires me!

One summer she was able to intern in our lab as a product safety test technician, and over the years I have supported her and her SWENext club in many endeavors. I also started giving my time to other organizations that promoted female engineers, and women and girls in STEM, because I knew firsthand how that this is a male-dominated field. I had experienced the value of having a female perspective in STEM. Women think differently than men; we come to conclusions in different ways; we see opportunities differently; and we conduct business in unique ways. All of these things add value! Engineering, business, and STEM as a whole would not reach the full potential if women were not a part of it.

Some of the highlights of my career are directly related to hearing the voice of God in my life. He speaks, I listen and act. One initiative I had the opportunity to champion was establishing the America's Women's Network (AWN) within our company. One day I had this divine thought, I was thinking, *I had been given this amazing opportunity that I only found out about because they saw me on an International Women in Engineering Day panel discussion that I was part of, but what about all the other ladies in our company? I'll bet they had never heard of this network and weren't given the opportunity like I was. So, what if we establish our own network here, kind of like extending the olive branch from an international network to a national network?* I reached out to the International Women's Network with the idea, and they were very supportive. They connected me with a female V.P. in the U.S. Day by day, this network started to develop. I developed a framework and the "why" behind this network. I also spoke with several ladies and requested their input. Soon a foundation was laid.

In April of 2019, I was invited to Arizona to attend a massive training event held by our company. Since we are a global company, this training is the one event where we would have the greatest number of colleagues in one location. On April 2, 2019, our CEO, along with the International Women's Network's founder and myself, rolled out the America's Women's Network. I was in a room I never imagined myself being in, with some of the industry's top technical talent, and some incredible men and women who were in support. WOW! I never saw this happening in my life. As the month went by, I knew that in order for this network to be of value, we needed a budget. I sought the help of my mentor and my career coach to compile a proposal to submit to our CEO for a budget for our network for that year. Keep in mind that I am an engineer; this was way out of my comfort zone and not like anything I had ever done in my life, but I just kept putting one foot in front of the other. Oftentimes I did it scared, but still, I did it.

The day came to present the budget proposal to our CEO, and I was so nervous. As the hour approached, I could hear my heart pounding in my chest; my palms were sweating, my stomach was turning, and everything in me wanted to back out, but I knew I had to do this. It wasn't just for me; this was for all the women who were a part of the network. I had rehearsed it so many times, and I had done my research to make sure I could answer any questions about my proposal. I was confident that I had done my part. Since I am based in San Diego, California, and our CEO is based out of our headquarters in Boston, Massachusetts, this request was done virtually.

As I entered the "room," my career coach was there, and our V.P. of H.R. was as well. Then in came the CEO. I was greeted and greeted back, thanked him for his time and consideration, and then started my presentation. All of twenty minutes later, I had his full support! I could feel his excitement for the network, and on the spot, he granted us the full request for that year! I was ecstatic! I knew this was the start of something extraordinary!

In the America's Women's Network, our mission is to shape the future of businesswomen in TÜV SÜD by providing opportunities for development, fostering valuable connections, and facilitating member success through business and career growth in a safe, sincere environment accessible to all. I have seen women within the organization be impacted in such a positive way by the America's Women's Network, which has grown to over 100 women in over 13 different offices throughout the U.S., Canada, Mexico, and Brazil. It blows my mind to see this go from nonexistent to something so valuable for the company. It adds an element of diversity we previously hadn't had and has been valuable for each woman involved. I tell you this to encourage you to step out of your comfort zone. You never know whose lives will be impacted by doing so.

Another highlight is being recognized by the San Diego Business Journal as one of the Top 50 Women of Influence in Engineering, for the last 2 years. I've had the pleasure of speaking at many STEM events, as well as be a guest on a number of podcasts and shows. My favorite part is meeting new people and supporting our next generation.

Over the years, I have partnered with some great organizations that have the same mission of promoting women and girls in STEM. I would attend or be a part of these events and walk away so inspired! In this book, you get to meet and be inspired by some of those ladies. At times, I would walk away from a STEM summit or conference thinking, *These women are extraordinary!* I can't believe some of the things they do, and some of the organizations they are a part of. The direct impact they have on the development of our society is tremendous. Being so inspired by what they do is part of what prompted me to write this book.

One day while running on the treadmill, I again had this divine thought, *I should put together a book on those extraordinary engineers. Those who share their amazing stories about what they do and how we are all impacted by them.* This was coupled with the fact that I had no idea about this amazing world of engineering until I was in my early twenties. There are so many cool women with careers in engineering that have incredible stories that need to be shared. My hope is that this will open you up to that world far earlier than me, and that it will give you greater insight into each discipline in engineering, what that career looks like, as well as what you can do with it.

So, what makes an engineer extraordinary? When I think of extraordinary, I think of someone who goes above and beyond. It's someone who does well in their responsibilities, but also does extra. Extra giving to something bigger than themselves. It is the person who doesn't settle for what has (or hasn't) been handed to them; they go after what they want. They are the changemakers, trailblazers, initiators, and creators. The ones who invest in and open up opportunities for others. It's truly an honor that these extraordinary women would be a part of this book, and I hope you are inspired and activated by hearing from them.

Connect with Justina:

QUESTIONS AND ANSWERS WITH THE EXTRAORDINARY ENGINEERS

CUENTAME CON SONIA

Contributed by Sonia Camacho:
Rotational Engineer at Nike®

Can you tell me about yourself?

My name is Sonia Camacho. I'm twenty-three years old and I work as an engineer at Nike. I received my Bachelor's of Science in Computer Science Systems degree from Oregon State University, and then I got a Master's of Engineering and Computer Science degree.

My journey into STEM (Science, Technology, Engineering, and Math) started back when I was in high school. I grew up with successful Latino parents, but my identity was something I struggled with a lot. Both of my parents have been my inspiration and helped me get to where I am today. While in high school, my mother was a big encouragement to me. As I mentioned, I struggled a lot with my identity and who I was, but I knew I always loved science and math. My mom was a math major in college and she always urged my sister and me to take those classes. But my sister and I are complete opposites! She is more reserved than I am. In high school, I enjoyed being a "girly girl" and was on the cheerleading team because I wanted to be in *"that"* scene. I remember picking out my classes, and when it came to choosing my electives, I wanted to take pottery with my friends. My mom said, "Maybe you should consider intro to engineering." I thought, *Oh, my gosh, no way. That's what my sister does. She's such a nerd. I would never do that.*

Why did you choose your field of engineering?

The reason I chose computer science systems, is because I loved the computer-aided drafting classes. I respected how those classes included, being able to draw on a computer and getting creative. I thought, *You know what? This is like a puzzle.* With good grades, I made it to AP computer science in high school, and things went even deeper.

I remember watching a video my teacher showed us about what computer science can do, and what fields are in it. There was every field under the sun. They mentioned fashion at one point, along with other things that you would never think would need computer science. I thought, *Wow, computer science is everywhere! You can do anything with it.* That video really solidified the decision for me. It was then, that I knew I was going to pick computer science for my career.

Where do you work and how did you get there?

I'm a rotational engineer at Nike. At such a big company, if you are a technology intern, you become a full-time intern for a year, and then you go into a "real" position. Right now, I'm trying different roles, but in about a year, I'll know exactly what I'm going to do.

In college, when you're a junior looking for an internship, whatever internship you land, is likely going to be your "full-time job" after college. When I was a junior, my internship was originally with Macy's (the department store). I was going to be on their technology team, but when the pandemic hit, that internship got canceled. I thought, *Oh, my gosh, what am I going to do? I'm going to be a senior with zero experience.* That is when I found out about the accelerated master's program at OSU.

In the master's program, I would stay in college after my senior year for one extra year. Afterward, I would earn both degrees. That following summer, I graduated with my bachelor's degree and I applied for the Nike internship. I thought being born and raised in Oregon and working for an Oregon-based company would be great! Plus, I knew I wouldn't be far from home. I loved the fact that I would use my degree to work with apparel, because, as I said, I was (and still am) obsessed with fashion and style! This is why I applied for the internship, and I got it! Initially, it was virtual, which was fine. But after a time, I went back to school for that last year to get my master's degree. Upon completion, I received the offer for the Rotational Engineer position. I am very blessed to have had this opportunity!

What are your responsibilities?

Right now, I am a Technical Product Manager. That means if there is a concept or a feature to be implemented, our team will work on that. I am a part of a "business side" team, and I work with the developers. On the business side, when the buyers say, "We're experiencing issues and they are impacting software," my team will research and find out what the issues are. We then go to the developers and let them know where there are issues with the software.

Right now, I am acting as a middleman between business and technology. I love this because I love to talk. It is a way to use my engineering skills, but be in a social setting. Because I have the experience from coding, I can say, "If you want that done, you're looking at a three-month timeline," or "We can get that done tomorrow." In my current role as a Rotational Engineer, I am in a lot of meetings. I must listen and understand different requirements, then be able to relay those requirements to different teams. It is great because we are all working to get the common goal completed.

Can you share some highlights of your career?

I've always been a mentor to other students on and off campus. During the pandemic I set up Zoom calls with different students throughout the country. It was then that I realized I could help and inspire others. My role has always been to be a "big sister" because I grew up with a big sister who is in STEM. I know not everyone has that, and I am so glad I did. Giving that level of help and mentorship is one of my biggest accomplishments.

Last year I went to the Society of Hispanic Professional Engineers (SHPE) National Convention where I worked with the SHPE National Team. That was one of the coolest moments ever! A lot of influencers do brand trips for makeup or product brands, and this was my version of that. I was able to go to a convention for engineers, along with other professionals, while also doing my social media work. That convention was one of those "pinch me" moments!! Social media, engineering, fashion—it all came together! When I think back to a few years ago, it wasn't like this at all. I would be in front of my laptop posting content and people would make fun of me. Now, I have worked with over fifty major brands. That is something I would have never dreamed of!

I was also recently flown out to Puerto Rico to be a judge at a national robotics competition. That was such a great experience! And most recently I started a podcast called *Cuéntame con Sonia Camacho*. In my podcast, I interview Latinos from different backgrounds and different fields. It has been amazing to be able to bring all my passions together!

What was more challenging than you had anticipated?

Time management! I think everyone would say that. It can be difficult when you are an ambitious person trying to be present on social media, working, and everything in between. I have learned that if you set priorities, you will make things work the way you truly want them to. It is important to know that no one is going to be there to hold your hand 100% of the time.

In high school you get assignments, and we are told exactly how to do them. We meet the criteria, turn in the assignments, and get a grade. In college, no one is going to care if you go to your lecture. As you get older, and you are working in the industry, things can be vague. There is not always going to be someone there to tell you exactly how to do things. That was a big wake up call for me, because I would get a project and think, I*s anyone going to tell me exactly how to do this?* The truth is, if there were someone to explain everything in detail, then they would most likely be doing it themselves. That's what I love about social media—we can share certain parts of what we do, and what we go through, hoping it will help someone else. As you mature, you learn these processes.

Another challenge I had, was dealing with imposter syndrome. On social media we all make it seem like we are confident all the time. But, there are days when you are sitting in meetings with a ton of people and you think, *I feel like the dumbest person here.* I used to feel that way in school, too. Thankfully I have gotten better at making that feeling go away. Imposter syndrome is something we need to realize will always be there, however, we need to know how to deal with it and not let it consume us. This is something I thought would completely go away after school, but it didn't, and that is okay. Now I know how to deal with it.

What do you wish you knew before choosing this career?

I wish I knew that it's not as difficult as the media makes it out to be. In movies, they have girls doing quantum physics because she's the "STEM girl." But, this is not brain surgery or neuroscience. You don't need to be the top mathematician to do computer science. Because of this, I think many people get discouraged, including me. I wish I knew sooner that not every engineer is perfect in every aspect! You can be a talented engineer, even if you struggle in certain areas. And it's okay to ask for help when you struggle. Having a growth mindset, means being okay with failure, because how will you grow if you don't fail? We tend to get really caught up in being perfect all of the time; I wish I knew earlier that I don't have to be perfect all the time.

Another thing I wish I knew sooner is that there are communities filled with people who are just like me. It took me almost two years to find a club at Oregon State for women in computer science. I always felt like the outsider that first year and a half, even making friendships that weren't great. Two years later, I found a community and my world changed. I got new roommates, and I had a support system. I had girlfriends to study with at the library. I never thought these kinds of things would happen, because we are such a small minority in the engineering field. The truth is, there are people who are for you—you just have to search for them.

Can you share some additional ways you promote women in STEM?

Giving back to my community is important to me, whether it's at Oregon State or mentoring (on or off campus). I support different club meetings by sharing my experience and giving advice to current students. There are questions sent to me through TikTok and Instagram, and I do my best to answer them. I share resources, whether they're for scholarships, programs, or internships. Right now, my social media content is going through a transitional phase. When I was a student, I shared study tips. Now that I'm in my adult life, I don't have finals, so I am not catering my content to being a student anymore. Instead, I share my "day in the life" activity hoping someone will see it and think, "Oh, that's cool! She looks like me, she acts like me, and she's an engineer at a dream company!" I've also been making my content educational while incorporating more about lifestyle.

Is there anything else you would like to share with the reader?

A piece of advice that I want to give, is that there is not a good time to start anything, so like Nike says, "Just do it!" If you get an idea, just do it. Don't be afraid of being bullied. I've had people send my content in group chats and make fun of me, but I am the one working with big brands at the end of the day. No matter what, I continue working towards my goals. Don't waste time and energy caring about what other people think; work towards your goals. Those who want to celebrate your accomplishments will find their way to you. If you have an idea, just do it! If you want to try STEM, JUST DO IT! There's never a right time. There's no waiting for next year. Just do it! Now!

Connect with Sonia:

@SONIA_MACHO

ASPIRING ENGINEER

Contributed by Madalyn Nguyen:
Aspiring Engineer at Worcester Polytechnic Institute (WPI)

Can you tell me about yourself?

My name is Madalyn Nguyen, and I was born into a first-generation family of refugees and immigrants. To avoid persecution by the Viet Cong, my mom and family escaped from Vietnam on a small fishing boat down the Mekong River during the fall of Saigon in 1975. Born in the United States almost thirty years later, I get to celebrate both cultures as my family blends American culture with Vietnamese culture. Although most of my relatives are healthcare professionals, I carved out my own path in engineering. After successfully participating in robotics and getting a certification from the engineering academy at my high school, I am pursuing a computer science major on a full merit scholarship at Worcester Polytechnic Institute (WPI).

Why did you choose the field of engineering?

Receiving a Mindstorm Robot in the fifth grade ignited my love for robotics and the desire to join every FIRST Robotics program possible: FIRST Lego League (sixth grade), FIRST Tech Challenge & SWENext (seventh grade), and FIRST Robotics Competition-Team Spyder (experienced/motivated eighth grader).

Since FIRST has given me the opportunity to hone my leadership skills and build my confidence and self-esteem, I wanted to "pay it forward" by helping other students in robotics, especially the underserved and underrepresented. As the daughter of a war refugee and immigrant, I have focused the past five years on two missions that are most important to me: making STEM accessible for everyone, and inspiring young kids to dream big and aim high, regardless of their background or income. Although it is a near impossible task, I have made it my mission to inspire as many people as possible.

What are you involved in at school?

I am serving on the WPI Executive Board as the Internal Events Coordinator. I am proud to be the only freshman to be elected to the board last fall. I am also an active member of SASE (Society of Asian Scientists and Engineers) and the Vietnamese Student Association.

I continue to be a STEM role model and mentor in my spare time.

Can you share some highlights of your STEM journey?

I am proud to say that I have reached the pinnacle in FIRST Robotics (FIRST Dean's List, which is the most prestigious individual award in FIRST), Girl Scouts-Gold Award, and the Society of Women Engineers-SWENext Global Innovator Award.

The initiative of which I am most proud is introducing the FIRST Robotics program to Wilson Middle School. Wilson Middle School is located in City Heights, California, which is known as a low socio-economic area with 100 percent of students receiving free or reduced lunches. Many of these students are immigrants or refugees. I knew it was a place where students didn't have all of the opportunities that I did to be exposed to STEM. It was also the middle school my mother attended when she first moved to California. I wanted to give these kids a chance to participate in robotics because I knew that even if the majority of the students didn't find it interesting, at least a couple of them would find their passion like I did.

Back in 2018, I ran a brief informational meeting at the school, where I was able to gain the interest of about ten young boys. It was barely enough to create a LEGO Robotics Team. Throughout their first year, I helped support them because I knew it would be difficult to keep the program running until it was sustainable. I helped get them the custom tables they needed to run their robot on and procured a

$1000 grant to fund their registration and pay for their robotics kits. I also helped them design a logo and get T-shirts for their team.

Just a year later, the club tripled in size, with one-third of the students being female. The club was able to create multiple teams and has won multiple awards. Unfortunately, due to the COVID-19 pandemic, I couldn't visit much throughout 2020, but I've recently been able to come back to do another hands-on mentoring session with all of the new students. The club has continued to grow and now requires tryouts to be part of it. It also has a waitlist of interested students.

During the pandemic, the club continued to flourish and even started an all-girls team. Again, I was able to procure another $500 Qualcomm grant for Wilson. In its fifth year, it finally has its first female club president. This year, going into its sixth year, I continue to support and mentor this program. Wilson Robotics is again thriving, as the seeds I had sown in 2018 are still flourishing as more underserved and underrepresented youth continue to have STEM opportunities from the robotics program. I created a legacy at this Title 1 school that will continue to impact hundreds and thousands of future underserved and underrepresented students.

What was more challenging than you had anticipated?

As an active participant and mentor in FIRST Robotics for the past seven years, I had the incredible opportunity to explore the world of STEM and really use my creativity to design and innovate. But being a BIPOC and a female wanting to pursue a male-dominated field that lacked diversity, had its road bumps.

Through this journey of discovery, I have experienced stereotypes and bias, and faced backlash and microaggression from peers and mentors while leading various robotics teams, groups, and projects. It was common to hear, "Are you sure you know what you're doing?" Or, "You're not ready for this." None of these challenges deterred me from pursuing my passion for STEM; rather, they encouraged and motivated me to go farther. I wonder if some of the doubters asked these questions because they've never worked with a woman or BIPOC on STEM projects.

In addition, I have encountered overt racism, such as high school boys making "chinky eyes" at me while calling me profane names, or more covert sexism, like trying to steer me away from the rigors of robot-building, due to implicit assumptions that girls have inferior technical skills. These things did not deter my STEM aspirations. I eventually led my team to build the robot as the chief project officer in charge of managing fifteen subsystems.

These experiences only fueled my mission to promote STEM diversity and help bridge the STEM gender gap. As a high school freshman, I founded two girls' engineering clubs (PHS SWENext and PHS NCWIT) to empower girls to pursue STEM. The impact is immeasurable, as the inaugural members are pursuing STEM careers and attending prestigious universities.

I have made it my mission to advocate for the underrepresented and less privileged, and provide them with the opportunity to excel in STEM without hesitation. When children are not exposed to STEM at a young age, they are less likely to believe that STEM is a viable career option for them. This is especially true for girls and students in low-socioeconomic communities. This is why I created the FIRST Robotics program at Wilson Middle School in San Diego, where the students are primarily minorities and/or children of immigrants and refugees. I have also mentored global robotics teams from Paraguay, Libya, and Benin. Additionally, I have conducted numerous outreaches to spread the message of FIRST Robotics to those who may feel that they don't have STEM role models, or that STEM isn't a place for them. As a mentor, I have been the keynote speaker at the Society of Women Engineers conference, and a speaker/panelist for other speaking engagements.

What do you wish you knew before you started your STEM journey?

Imposter syndrome is real. Even in college, I face it frequently and as often as weekly. As a FIRST Community Scholar on a WPI full merit scholarship, some boys at WPI would deem me not worthy of the award, simply because I am a female.

Belonging to the Society of Women Engineers (SWE) on any college campus is the ultimate way to combat imposter syndrome. It's important to belong to a like-minded community to build a network of support.

Can you share additional ways you promote women in STEM?

I recall my heart racing, shortness of breath, clouded vision, and beads of sweat flowing down my neck as I stood on stage with what seemed like hundreds of eyes tracking my every move. Previously, I had dreaded speaking in front of the class or even raising my hand. But since then, I had found myself on stage as an eighth grader talking about FIRST Robotics to inspire students to participate, while simultaneously experiencing my first panic attack, unbeknownst to the audience.

As a child, I loved tinkering with things. LEGOs were my bread and butter, and every household object that could come apart in my house was another challenge to reverse engineer. This interest took me to FIRST robotics in the sixth grade, where I experimented with my curiosity and developed a love for innovating. I had a vision that all students should have the same opportunity that I did, which is why I nervously accepted an invitation from my middle school to speak and encourage students. I presented, held a Q&A, and survived a panic attack while speaking. This experience, while nerve-wracking, taught me a valuable lesson about overcoming my fears.

What I once feared is now my greatest skill, joy, and tool to advocate for STEM education and diversity. I love speaking about STEM to anyone willing to listen. Practice makes one perfect, so I continue to hone my oratory communication skills. Nowadays, you can find me talking to young kids, small audiences, and large audiences about STEM. I have spoken at many events, such as a SWE school luncheon (500 middle schoolers), board of education, city council meetings, and various panels. As a robotics copresident, I was often the spokesperson for my team on the news or presenting to award judges.

Fast forward five years, and that scared eighth grader who had a panic attack on stage is now able to be a keynote speaker for the SWE conference, an emcee, and an event moderator. Public speaking was my insecurity, and now it is my greatest strength to help encourage children and peers toward the mission of advancing global ingenuity.

Is there anything else you would like to share with the reader?

As the Chief Project Officer (2020–2021) and Vice President of Public Relations (2019–2020) of my robotics team (Spyder 1622), I managed fifteen sub-teams to build and program an industrial-size robot for competition. I also organized informational nights for elementary and middle schools to create teams and to talk about robotics.

During the pandemic, as the Founder and President of the Poway High School Society of Women Engineers Club, I organized and moderated a virtual STEM Careers Series. Members of my clubs got a chance to cohost so they could experience public-speaking. We impacted over 700 attendees (students, parents, mentors, and SWE professionals). We hosted seven weekly Zoom sessions during May, June, and July 2020. For these efforts, we were recognized by the Society of Women Engineers for the WE Local 2021 Outstanding Outreach Event Award.

I successfully developed and organized the Global Engineering Challenge 2020, for which over 65 teams registered. The goal of this challenge is to give students STEM opportunities by using household materials during the pandemic. Overall, it was a huge success, with students from the second grade to the twelfth grade participating. A team in Paraguay won!

I also spearheaded the passing of a Women in STEM Day in the Poway Unified School District to promote diversity, equity, and inclusion. I organized a two-day Women in STEM forum, which was recognized by SWE as the WE Local 2022 Outstanding Outreach Event Award.

I helped organize the 2022 Girls in STEM Inspiration Day, and was recognized by SWE for the WE Local 2023 Outstanding Outreach Event Award. Girls In STEM (GIS) was originally organized in 2019 as my Gold Award Girl Scouts project, and was funded by the SWENext DesignLab Community Engagement Challenge. I started this project in 2019, and it was carried on by the Society of Women Engineers-San Diego Professional Section in April 2023. We have also hosted five Girls in STEM Camps, which impacted over 150 girls all over the USA.

Aspiring Engineer 135

Connect with Madalyn:

ADASTRASU

Contributed by Susan Martinez:
Mechanical Engineer in the space industry

Can you tell me about yourself?

My name is Susan Martinez. I am twenty-six years old and I have a Bachelor's Degree in Mechanical Engineering from the University of Kentucky. I graduated in May 2019 and I am married (since May 2019). I have an identical twin sister (also a mechanical engineer). I have a dog (Jupiter) and two cats (Juno and Ceres).

I was homeschooled for all of my schooling until college, when I got an Associates of Science from Ashland Community & Technical College (ACTC in Ashland, KY). I then started my internship at NASA Marshall Space Flight Center (MSFC) in the summer of 2016, in Centennial Challenges as an intern. I returned to school, but started at the University of Kentucky in the fall of 2016. I applied and was selected for a NASA Pathways Co-op at NASA MSFC as an engineer in the additive manufacturing group. I was a pathways co-op from 2017 to 2018. I then returned to school to graduate with my bachelor's degree and started at NASA in 2019 as a civil servant in the additive manufacturing group. I spent two years working full-time as an additive manufacturing engineer, supporting the SLS and RS-25 engines, and designing and producing test articles across NASA's needs.

I moved to become an operations controller in May 2021 for the International Space Station; this is a flight controller position supporting payload operations, safety, integration, and day-to-day operations on the ISS. In September 2022, I started at Blue Origin as a payload engineer for Orbital Reef.

Why did you choose your field of engineering?

I actually wanted to go into aerospace engineering, but at the time the University of Kentucky didn't offer that program as a degree, so I went into mechanical engineering. I knew a lot of mechanical engineers at NASA—more than aerospace engineers, in fact. I did get an "emphasis" in aerospace engineering with the electives that

I selected. I am now very thankful that I pursued mechanical over aerospace. I think it gave me a much broader education in a general engineering sense. Looking back, sometimes I wish I could have pursued systems engineering, as I am now working as a systems engineer. I am hoping to pursue a Master's of Systems Engineering sometime in the future.

What is your occupation and can you describe your journey to get there?

I work at Blue Origin. I'm a former NASA engineer! I always thought I would spend my entire career at NASA; you know, one of those old Apollo engineers that always says things like, "Well, back in the Apollo (or Shuttle!) days . . ." and that's what I wanted out of my career.

As I pursued more career advancement and changed roles, I found that I didn't really mesh well with the lifestyle of a flight controller (having to support the console 24/7, long hours to prepare, high-stress situations, etc.). This to no fault of anyone; it just did not suit me. I found that the longer I pressured myself to become a flight controller, the more unhappy I became. This led to me start looking for other opportunities within NASA, and also externally to other aerospace companies. I knew that this could potentially mean I had to give up my civil service position at NASA. The more I looked at NASA, the more it became obvious that my place was no longer at NASA. I applied to a lot of jobs, but once I applied for the payload engineer job at Blue Origin, everything started to make sense and fall into place. Since then, I haven't looked back. It has been a struggle fighting with the belief that I would spend my whole career within NASA. I left shortly after I got my five-year service pin.

What are your responsibilities?

I am a Payload RE for Blue's lunar program. I am developing payload strategies and payload integration and I will be working with NASA through Blue with the newly won SLS (Space Launch Systems) Sustaining Lunar Development contract. I am working on payload development from the ground up with requirements, and designing exactly how the payloads interact with the new station and the future astronauts. It involves a lot of research, engineering, and critical thinking, which is really what engineering is all about.

Can you share some highlights of your career?

I have been able to work on so many incredible things in my career so far that it's hard to pick a highlight. Starting in school, my senior design project was a robot (whose name was Oscar) for the NASA Robotic Mining Competition. My team and I were told early on that there were five previous years of senior design students who did not go to the RMC competition, and my professor was sick of it. He told us that we HAD to go to the competition, or we were all getting two letter grades less than what we deserved (I'm sure this was an empty threat, but we were not about to find out). We worked ourselves to death to design and build a robot (all by ourselves, might I add—all of the welding, machining, everything we did ourselves), and we went to the competition. It is truly one of the things I am most proud of in my engineering career.

I had so many days of work and accomplishments that I was proud of when I was an additive engineer, and especially when I was a flight controller. In relation to STEM (Science, Technology, Engineering, and Math), I have had many articles, features, speeches, and accomplishments, but my favorite is that I was named *SHE*'s STEMINIST of the Year for 2022.

While I was at the Reinvented Space Gala at Kennedy Space Center in 2021, a young woman came up to me and said she started pursuing engineering in college because of me and my outreach on Instagram, and that was worth more than any feature, speech, article, or anything else. That is the whole reason I got into STEM outreach in the first place; so that girls like me have a role model to look up to in engineering when I didn't. I believe the biggest highlights of my engineering career, STEM outreach, and science communication career are yet to come.

What was more challenging than you had anticipated?

Leaving NASA. It seems silly, but that to me was always the "end-all, be-all," and it just... is another engineering job. As hard as that reality was to accept, it was one of those things where I accomplished my GOAL, my real-life goal, and it was like, "Okay, now what? Is this it?" I had to take a long, hard look at what I actually wanted to accomplish with my career rather than wanting to work somewhere because the name was "NASA," or wherever your dream place to work is. I realized that I really wanted to help advance space flight and living in space, and be a part of changing the narrative for women in STEM. I realized I was not in a position to do that at NASA. It was a harsh reality, but one that I believe turned out for the best.

What do you wish you knew before choosing this career?

That it's never going to be easy. School was difficult for me. No one is going to take you seriously until you "prove" yourself. This is frustrating when you feel like others didn't have to prove themselves quite as much as you did. My mom always said, "Girls have to work twice as hard to be half as good." I always worked 10 times as hard just to be on the same playing field, and I think it really has helped advance my career. It created this crazy work ethic that I have to be the "best" that I can be, whatever that looks like.

What are some additional ways you promote women in STEM?

Instagram is a huge platform for me. I do at least 90 percent of my outreach through Instagram. I have done conferences, keynote speeches, featured speaking, etc. I do a lot of mentoring through Instagram DMs; I have helped quite a few students with applications, resumes, and internships. It is a pretty unconventional way, but it gets the job done.

Is there anything else you would like to share with the reader?

To my women in STEM, to girls thinking about pursuing STEM, and to those scared of the daunting task of a degree in a STEM field: we need YOU. We need your knowledge, your experience, your mindset, and the way YOU think through problems. If you are considering pursuing STEM, there is a place for you. No one will be you; you are the only person who can bring your uniqueness to the job, the company, the school, the group.

Connect with Susie:

REINVENTED

Contributed by Caeley Looney:
Aerospace Engineer & CEO

Can you tell me about yourself?

Hi! My name is Caeley Looney, and I'm an aerospace engineer. I grew up on Long Island, New York, and consider myself *incredibly* lucky because both my parents were engineers. My mom, in particular, was a naval engineer, so our house was constantly filled with blueprints of aircraft carriers and other really cool engineering schematics. Naturally, I decided I wanted to be a fashion designer. However, if you saw my style back then, you'd quickly realize that the world wasn't ready for my wacky sense of fashion, and it was probably for the best that my mom helped me discover my other passions.

When I was in the sixth grade, my mom saw an ad in our Girl Scout program book for FIRST Robotics. She pretty much had to drag me to this informational meeting, because at the time I had absolutely no interest in STEM (Science, Technology, Engineering, and Math). It was something I was good at in school, but never really considered it to be a passion of mine. But when I saw the robots in action, I completely fell in love. I think it kind of made me realize that STEM was so much more than what I had experienced in the classroom, and that the real-world applications of what I was learning were so much more intriguing and hands-on than what I had read in my textbooks. Learning about robots is one thing; seeing robots in person and playing with robotics kits is completely different. It sparked that lifelong curiosity for me.

Why did you choose your field of engineering?

I am jumping around a bit here in my story, but something that is super important to me is that I'm a Latina. My mom was born in Ecuador and moved here with her mom and sister when she was young. She chose engineering because, at the time, marine biology, which was her passion, wasn't a career that she felt was realistic for her. I ended up choosing aerospace engineering, but it wasn't because it was a logical and safe career choice; I chose it out of passion. I owe a lot of that to my mom because she constantly pushed me to go for my dreams, rather than feel confined to a box that society often places young girls in.

I honestly think I kind of stumbled into aerospace engineering because, at the end of the day, I really just thought space was pretty cool. I took a research class in high school where each semester we worked on our own individual projects. I didn't really know what I wanted to do, so my teacher suggested that I start with something I already liked; that was robotics. My first project was to build an underwater ROV (remotely operated vehicle), or pretty much a robot that worked underwater, and I HATED it. I liked the building process, but the idea of working underwater was just something that didn't spark much joy for me. Then my teacher suggested the complete opposite. He told me to go learn more about robotic applications in space. This got my brain excited. It completely blew my mind that robots not only existed in space, but were also doing research in space! I learned more about the latest Mars rover, and the rovers that launched and landed before that. It was then that I decided that I was really curious about technology and its applications to space exploration and space travel. Therefore, when I went to college, I chose to pursue a Bachelor's of Science, in aerospace engineering.

It wasn't until my last semester of college that I really honed in on which part of aerospace engineering that I specifically wanted to work in. I interned at Kennedy Space Center in the fall of 2017, right before I graduated. I had the opportunity to work on a guidance, navigation, and control research project with engineers in their Swamp Works Lab. I learned all about something called dual quaternions and how we use code, algorithms, and software developed to help satellites and other spacecraft navigate in space. Now, I work a lot with GNC for small satellites in my full-time job.

What are your responsibilities in your current role?

Right now, I work primarily with small satellites, specifically on mission analysis. This means that I design the way our satellites will move and operate once they are in space. I determine what the orbit for each satellite will look like and help our subsystem engineers design their specific subsystems (i.e., propulsion, ADCS, power, communications, etc.).

In addition to this, I write flight software for our satellites. This flight software is used to help satellites avoid potential collisions with orbital debris or other spacecrafts that they might fly by. It is also used to help predict where our satellites will be over the course of their future orbits, so our ground team can coordinate when we can perform our missions, or when we can send new commands from our ground stations.

What was more challenging than you had anticipated?

I could tell you what is obvious: being a woman in STEM is not always easy. Especially in aerospace; an industry that is very much still male-dominated. It was hard for me to gain respect from my older, male coworkers when I started my first full-time job. I constantly felt like I had to do more and show more work, or I had to get mentors to sign off on my work in order for my managers or other engineers to accept my calculations or code. This definitely got easier the more that my coworkers saw that I was providing valuable insight into our satellite designs, but it was definitely still really hard on me at first.

The other thing I want to mention is that math and science are hard. It's not hard for everyone, but it definitely was for me. I failed multiple physics classes in college, but I still made it to where I am today. Engineering classes are difficult, and that can be really frustrating and discouraging, but if you're really excited and passionate about STEM, just keep at it and don't let a failing grade define you or your future!

What do you wish you knew before choosing this career?

I definitely wish I knew how much creativity went into aerospace engineering! It's really easy to get caught up in grades and exams in college and lose sight of the fact that engineering is actually all about innovation and creativity. On an exam, there is usually just one right answer. As students, we are taught only one way to do things, and we feel like that's the only way to do things. But in engineering, that is the farthest thing from the truth. And it's not only that there isn't just one right answer, but as engineers, we are actually encouraged to think differently and think of new ways to get to the same result! Remember when I said that I wanted to be a fashion designer originally? I've always loved art and being creative, and I gave that piece of me up when I left for college to become an engineer. But working in the field and seeing creativity encouraged has helped me find my own sense of creativity again!

Can you share additional ways you promote women in STEM?

In 2019, I started a nonprofit organization called Reinvented Inc., which aims to empower and inspire the next generation of girls in STEM. As the CEO and founder of the nonprofit, I have the incredible opportunity to work directly with girls all over the world to help spark in them, the same passion and curiosity for STEM that I found in middle school! We publish a print magazine featuring the stories of other professional women in STEM each quarter, and also have our Princesses with Powertools Program, which is my favorite! I get to dress up like a princess on the weekends and travel all over the U.S. teaching girls how to use their first power tools.

Is there anything else you would like to share with the reader?

I love making! Over the past year, I've really gotten into working on crazy projects on the weekends. My favorites to date have been a hand-forged sword, an origami kayak, an Olivia Rodrigo-themed smart mirror, and a 3D printed light-up constellation dress! I document the build process for each of my projects on my Instagram page in hopes to inspire and encourage others to pursue their own crazy maker shenanigans. Know that no idea is too crazy!

I also wanted to mention that I am on the autism spectrum and have ADHD. Rather than letting these disabilities hold me back, I let my mind (which works a bit differently than most) run free. This allows me to be really creative, not just at work, but in coming up with all of my crazy maker project ideas!

Reinvented 151

Connect with Caeley:

NOIRESTEMINIST®

Contributed by Dr. Carlotta Berry:
Electrical Engineer with a special in controls and robotics
at Rose-Hulman Institute of Technology

Photo courtesy of Rose-Hulman Institute of Technology/Bryan Cantwell.

Can you tell me about yourself?

I was born in Nashville, Tennessee, to a single mother and was the youngest and only girl. I had two older brothers; one is now a graphic artist, and the other does CAD work. My mom was a kindergarten teacher for over thirty years, and my grandmother was a piano teacher for thirty years. So teaching was in my blood, and this is all I ever wanted to be: a teacher. I loved to hold school with my dolls and assign them homework and grade it. I always loved math and grew to love science later.

Due to my affinity for learning, I was encouraged to attend a magnet school in the eighth grade. It was in high school, while I was in a program called Inroads, that I was encouraged to consider engineering as a career. I did not know what engineering was, and thought it was a train conductor. Therefore, I went to the library after school and read about it. It looked interesting, but I was still not sure. I did not know anyone who was an engineer. I always knew I was going to Spelman College, but in order to be safe I decided to pursue a dual degree between Spelman College in math and Georgia Tech in electrical engineering. This way I could pursue both of my passions. Since I always wanted to be a high school math teacher, I thought if engineering did not work out, I could return to my desire to be a math teacher. I think I selected electrical engineering because it was so math-intensive that it satisfied my first love. It would be later that I would select controls and then robotics as my chosen sub-discipline in engineering.

What is your current role and what are your responsibilities?

Initially, I worked as a controls engineer in a manufacturing plant for Ford Motor Company's Glass Division in Dearborn River Rouge, Michigan. I helped with maintenance on the assembly lines, debugging and troubleshooting when things went wrong, and programming the robots and logic controller computers, which controlled the assembly line that produced the car windshield and added the mirror button.

After that, I returned to graduate school at Wayne State to get a master's degree in controls, and then got a job working at Detroit Edison as a controls engineer. Detroit Edison was the electric utility company. After one year, I quit my job to enter graduate school to pursue a Ph.D. in electrical engineering, with a focus on robotics. I had decided while I was a student at Georgia Tech that I wanted to be an engineering educator because there was a severe lack of diversity in that field. I wanted to show that teaching engineering could be done in a different way which was diverse, inclusive, fun, and welcoming.

What was more challenging than you had anticipated?

I think that I never realized how bad the lack of diversity in STEM (Science, Technology, Engineering, and Math) was. I did not know how unwelcoming the field was to women, brown, and black people. I think maybe if I had been more aware of the bias, I would not have kept going when professors at Georgia Tech treated me poorly or administrators told me to quit and drop out of engineering school because of my grades, rather than considering that I worked a lot because I was out of financial aid and needed to eat.

It was equally as bad being an entry-level engineer in a manufacturing facility, where I experienced racism, ageism, and sexism. Many times, I continued moving forward toward my goals despite the obstacle course, without being equipped with the knowledge and resources to be successful. If not for the supportive environment of Spelman College, a historically Black women's college that built me, encouraged me, helped establish my self-esteem, and helped me see my purpose and vision for life, I may not have been able to endure all of these unanticipated challenges of a career in engineering. If there had been no Spelman College, I would not have graduated from Georgia Tech, Wayne State, Vanderbilt, or become an engineering educator. I actually modeled my value as an engineering educator after the way that I was taught, mentored, and supported at Spelman

College. Although I had seen Black women get Ph.D.'s, I had never seen one in engineering. Mae Jemison was probably the closest I had seen to a role model for where I aspired to go. I would later learn more and find mentors in my field.

What do you wish you knew before choosing this career?

Thinking about the previous response to the question, I don't think I needed to know about the difficulties of the obstacle course in advance, because I may have turned back. However, now that I do, I use what I have learned to pay it forward by being a mentor to K–12, undergraduate, graduate students, and faculty. I worked with a colleague to establish the ROSE-BUD Rose Building Undergraduate Diversity program, which is a scholarship, networking, and professional development program for women, marginalized, and minoritized students to pursue, develop, and excel in computer science, computer, electrical, and software engineering.

Can you share additional ways you promote STEM?

In addition to the other activities I mentioned, I am the cofounder of Black in Engineering and Black in Robotics. We teach a wider swath of the community about robotics and STEM, promote anti-racism in engineering education, and curate content on a YouTube channel and social media to teach about STEM and robotics. I use my fictional books about Black women in STEM and Robot Slam Poetry to teach kids and adults about STEM and robotics concepts in small snippets. Also, I am an open-source hardware trailblazer fellow, designing open-source mobile robots to guide academics in engaging in open-source robotics for service, teaching, and research. I live by the motto, "My STEM is for the streets," to show others diversity in STEM and encourage them to join me in making STEM more diverse, so we can design the solutions and processes to improve the world. We can remove injustice and bias in STEM education and technology.

Connect with Dr. Carlotta:

"Trust in the Lord with all thine heart, lean not to thy own understanding. In all thy ways acknowledge him and he shall direct thy paths."

(Proverbs 3:5–6)

Photo courtesy of Rose-Hulman Institute of Technology/Bryan Cantwell.

WATCH ME STEM

Contributed by Paula Garcia Todd: Chemical Engineer at IFF
(International Flavors & Fragrances)

Can you tell me about yourself?

Hi! I'm Paula Garcia Todd. I am originally from Brazil. My parents are from Chile and moved to Brazil before I was born. We moved to the States when I was about ten years old. I didn't speak a word of English when we moved here, and I was never sure where I really belonged. I didn't feel Brazilian enough because my family was not originally from there. I didn't feel Chilean enough because I had never lived there, although I often visited my extended family in Santiago. I obviously didn't feel American at all when I was young, but I do very much now as an adult. I'm surprised by the number of people that can relate to those feelings when I share them. Your uniqueness is what makes you special. You bring a perspective of mixed experiences and backgrounds that nobody else has—remember this important point.

Why did you choose your field of engineering?

I was lucky to be born into a family of engineers. My grandfather, father, uncles, and brothers are all engineers. I grew up in an environment where I was never asked what I wanted to be when I grew up, but rather what types of problems I wanted to solve. When you're young, you usually consider careers that are evident around you: teachers, firefighters, and doctors. I thought I wanted to be a doctor when I was little, but I literally fainted at the sight of blood (you can ask my three kids: when they get injured, they go straight to Dad!). Medicine clearly wasn't in the cards for me, so the next best step in my mind was working in the pharmaceutical field. I chose to study chemical engineering at Penn State University in order to have a broad background that would allow me to do various types of jobs in the pharmaceutical field. This has proven true for my nineteen-year career! As I expected, engineering allowed me to touch the medical field in a different way than a doctor would. Engineers are behind so many fields and industries. They impact the progress of new technologies across the board.

What is your current role?

I have been working with inactive ingredients that go into pharmaceuticals for the majority of my career. These ingredients are important to ensure that medicines are effective in your body, and that pills and capsules are made and processed correctly. I love that the chemicals I work with come from trees and seaweed—natural products that we use in a sustainable way! I started my career as an engineer in a chemical plant; it was cool to see how things are made. You also get to climb and fix huge equipment in a chemical plant! I then spent some time in the lab, creating new molecules and products.

Describe the journey to get to where you are?

About nine years ago, I decided to try something completely new and took a job in marketing; this led to other business and strategy roles. I had never taken a business class in my life, but the fact that I knew how to solve problems and had a solid technical understanding of how pharmaceutical scientists use our products, made me perfect for these jobs. Even in these roles, I utilized my technical knowledge every day to be successful. I've had many friends with engineering backgrounds go into a variety of fields and career tracks—you can make what you wish out of your career; it never gets boring. Recently, I accepted a new job as the leader of our innovation group. I lead a group of forty scientists and engineers to develop new products for the pharmaceutical industry.

What are the responsibilities in your current role?

I lead a team of amazing scientists to help pharmaceutical scientists ensure their medicines are working correctly in the human body. Our team is global, having representation in Germany, India, China, and the U.S. We use chemistry and polymer science to develop new innovations that solve critical challenges in getting drugs delivered and absorbed into the right parts of the body. Our products are found in one-third of oral medicines across the globe! Our team gives presentations at tradeshows and customer sites, performs experiments to help customers develop new application areas, advances new products based on customer needs, and advocates for customer needs internally across our business.

Can you share some highlights of your career?

I have had the opportunity to experience so many amazing things throughout my career. I've had the pleasure of collaborating with scientists from pharmaceutical companies around the globe to help develop their products. This has allowed me to visit many countries and experience many cultures. I have been able to file patents on both new polymer discoveries I have made, and brand-new products that I have created and brought to market. I have held a variety of jobs that have allowed me to have an in depth understanding of how labs and manufacturing settings work, as well as teaching me how to run a business. I have been asked to speak in many settings, from small customer venues to large corporate settings, in front of investors, and other 176 business leaders.

Lastly, I've had the gift of being able to do a job that I know impacts people around the world every single day. I know I enable drug delivery technologies to work, therefore allowing patients to get healthy more quickly and more efficiently. There is no greater accomplishment than knowing that my engineering background is helping people!

What do you wish you knew before choosing this career?

Engineering can be hard, but just because something feels hard it doesn't mean you weren't meant to do it. In fact, when something is hard or requires practice, that's when real learning takes place.

Challenges should not make you shy away; they should keep you curious and persistent. Give yourself grace if you need a break, but then come right back to that thing that felt hard, and you'll continue to grow beyond your dreams.

I also wish someone had told me sooner to find a community for support. When you surround yourself with people who care about you, you all push each other forward much further. Connect with other girls in your math and engineering classes, and cheer each other on. Learn from each other, study with each other, and lean on each other—you will all be better for it!

Can you share additional ways you promote women in STEM?

I've learned about the social impact you can make as an engineer, not just in your job but in your community as well. I enjoy spending time visiting classrooms to normalize the concept of women in STEM (Science, Technology, Engineering, and Math) fields—to help inspire young girls and show all kids that STEM is for everyone. I especially love sharing my experiences and mentoring students that are underrepresented in STEM, because diversity is crucial for innovation. In that service work, not only have I felt personally fulfilled, but I've also received some recognition that I'm very proud of.

The American Association for the Advancement of Science named me an IF/THEN ambassador. I received my own life-sized 3D printed statue that was exhibited at the Smithsonian Museum in Washington, D.C.! I also got to do an episode of the CBS TV show *Mission Unstoppable,* and was named Woman of the Year by Women in Technology in 2020.

Having given interviews for newspapers and magazines, spoken at museums, been on STEM podcasts, and mentored educators—all in the name of helping more people see the beauty and benefits of bringing more girls and Latinas like myself into STEM fields! Please realize that you—yes, YOU—have the power to be a mentor and role model to someone else, and you can use that power as early as NOW.

Is there anything else you would like to share with the reader?

By the time you graduate, there will be jobs available that don't even exist today! Don't worry too much about what specific job you want to do when you grow up; think about the problems you want to fix to help guide you towards the industries that are interesting to you. Finding a fun job will come easily after that. Follow your dreams—and your talents—and pay attention to things you are good at, as this will help you discern what you might enjoy doing in your career.

Connect with Paula:

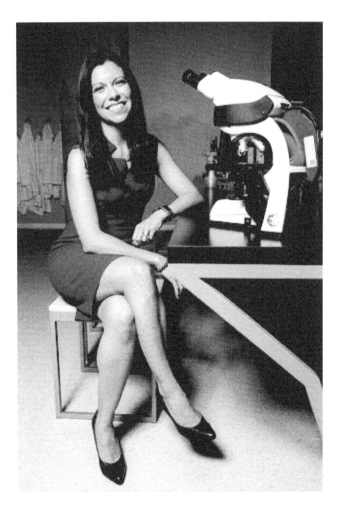

ALL THINGS CHEMICAL ENGINEERING

Contributed by Alexis Enacopol:
Validation Engineer

Can you tell me about yourself?

My name is Alexis, and I'm a chemical engineer working in the medical device and pharmaceutical industries. More specifically, I work as a validation engineer in a variety of areas, from validating manufacturing processes to production equipment. While earning my degree in chemical engineering, I was an undergraduate researcher in a metabolic engineering research laboratory within the School of Engineering at my university. My involvement in this research group concluded with a publication titled "In vivo production of psilocybin in E. coli."

Why did you choose your field of engineering?

I chose to major in chemical engineering because I've always been interested in chemistry, but I knew this degree would allow me to pursue many different fields, from the medical device/pharmaceutical industry, to the cosmetic industry, to the petroleum industry. This career flexibility was something that attracted me toward this choice as I was in the decision-making process. It can be a bit overwhelming to decide what you'd like to do for (most likely) the rest of your life at the juvenile age of eighteen, but chemical engineering alleviated that stress a bit when I understood how many paths I could take with this degree.

Another aspect that had some role to play in declaring my major was the fact that, from a young age, I'd always wanted to be involved in the medical field in some way. I wasn't sure how, but it never crossed my mind that I wouldn't end up working in that industry somehow. I knew that by getting involved in the healthcare field, I would be able to see the impact I have on patients who use the medical devices that I help bring to market. One of my favorite parts of my role is being part of the production process of many medical supplies, from those that can be routinely seen in doctor's offices to more involved devices such as humidifier components. Even if the product doesn't affect my day-to-day life, I know it impacts someone else's life. For example, the humidifiers I previously mentioned were in high demand for COVID-19 patients early in the pandemic.

Can you describe the journey to get to your current role?

I work in medical device and pharmaceutical manufacturing. My first role was in equipment validation at a CDMO (contract development manufacturing organization) facility, where I learned the basics of validation and what a validation project entails. A contract development and manufacturing facility is a company that conducts drug development and manufacturing for a separate company. The CDMO facility already has all the equipment, space, and resources required to produce the product in question, so it helps make the manufacturing process more efficient. After my first role focused on equipment validation, I moved onto process validations in a manufacturing setting.

My first full-time position was in a fast-paced consulting role, which allowed me to quickly gain validation experience and industry knowledge. I'm grateful I had the opportunity to work onsite for my first few months in validation, because I learned a lot of technical material that I was able to leverage while interviewing for my next role.

What are the responsibilities in your current position?

In the medical device and pharmaceutical industries, there are regulatory organizations or "bodies" that set requirements for testing and inspections which need to be met during the production of medical devices and medications. These requirements are set to ensure each product that leaves the doors of the manufacturing facility is exactly as intended—whether it's how it looks, how it functions, or how it was made. We do this to make sure people aren't using defective products, which could be life-threatening, depending on the device or medication in question. As a validation engineer, my responsibility is to ensure the manufacturing processes and equipment used meet all requirements established by these regulatory bodies.

Validation can be broken down into two main areas: document drafting and execution. Let's begin with document drafting. A validation study requires a protocol to be drafted, which explains all of the requirements and testing within the scope of this project. It states what the acceptance criteria of those requirements/tests are, and provides general background information on how different aspects of the validation strategy were determined and justified. Once the study is completed, a final report is drafted to summarize the findings.

Execution is the process of conducting the testing or running the manufacturing process per the established protocol. At many sites, operators who are trained on the piece of equipment, and work on it daily, conduct the execution portion of a validation study. Sometimes (similar to what I was doing in my first equipment validation role), the validation engineer may be the one going into the cleanrooms and managing the process. This depends on the facility and standard operating procedures (SOPs) at that facility.

Can you share some of the highlights of your career?

One of my favorite projects I worked on in my career, thus far, is validating bioreactors used for producing a COVID-19 vaccine. It was exhilarating to know how much of an impact the project had on the world. It was overwhelming at times when thinking about the ramifications of that impact if the project's expectations weren't met. I mentioned earlier that I wanted to see the impact I was making as I worked on validation projects, and it was clear as day how important this project was, given that it was taking place during the peak of the COVID-19 pandemic. It was stressful knowing that if deliverable deadlines weren't met for this project, millions of dollars' worth of vaccines would be lost, and huge quantities of the vaccine would not be able to be released to people who needed it at that very moment. I am truly grateful that I had the opportunity to work on such a significant project, and I still admire the diligent team I worked with to get it done.

One highlight of my day-to-day responsibilities is that I am able to work on a variety of problems, instead of finding myself working on routine and repetitive tasks. Sometimes, these issues might cause a product to be recalled or removed from the market. There are many consequences should this happen. Primarily, patients' lives are put at risk if the product is defective or not manufactured well. These types of situations cause a little bit more of an adrenaline rush as my team quickly works on understanding what went wrong, why it happened, what the impact on the product is, and the best way to move forward.

Another highlight of my career that I didn't expect when I started my role, was the fact that I would have the opportunity to travel as an engineer. Throughout college, I was always under the impression that the luxury of work travel was only enjoyed by people who pursued a business consulting role, solely because that's what was talked about at career fairs and other networking events. I had never heard of engineers going to my universities to discuss how they were traveling from site to site to work on different projects. I always thought each facility had engineers that solely specialized in their machines and processes. That thought was quickly proven false as I found myself traveling weekly to the East Coast from the Midwest to work on the COVID-19 initiative I highlighted earlier. In my following role, I still found myself traveling to various locations and visiting various manufacturing facilities to take a firsthand look at their equipment. I was surprised to realize that I would have the opportunity to travel with a chemical engineering background, and that was an appreciable realization to me. I look forward to seeing where else my career takes me (literally).

Another thing that seemed impossible to me but was a pleasant discovery was being able to work remotely as an engineer. Of course, the COVID-19 pandemic played a large part in changing these practices and making remote work a plausible option for engineers, as well as for other roles across many industries. Working remotely as an engineer quickly drove me to focus on developing and improving my communication skills when talking to colleagues via email. When you're on the manufacturing floor, it's easy to point to the root cause of an issue on a machine. How can someone eloquently get a point across when there is nothing tangible to refer to?

The last highlight of my career that I want to discuss is the type of people I work with. I have already touched on how diligent and perseverant my team was while working tirelessly on the COVID-19 vaccine initiative in my first role, but I want to emphasize how much I truly learned from every individual who was on my team. Each person excelled in a different area, and it was amazing to see all these strengths come together to successfully pull off the project. I hold high regard for my first team, and I continue to work with individuals daily that continuously challenge me to think about ideas from a different perspective. I value the fact that each day I can continue learning from others to build up my industry knowledge and skill set.

What was more challenging than you had anticipated?

One challenge I hadn't anticipated coming into the industry was how much collaboration is needed between multiple teams, especially in a validation role. To name just a few teams I work alongside, I collaborate with the engineering team to understand the ins and outs of the new equipment and processes I validate; the laboratory team to learn the background of certain tests and why they need to be conducted; and the manufacturing team to schedule validation studies during routine production runs. Each team has its own priorities for projects and tasks that need to be completed, which occasionally don't align with validation priorities. It was quite intimidating the first time I reached out to the head of the production team, a tenured employee and many years my senior, and requested production be halted for a high-priority validation run. If delayed, would've halted the sale of product, causing thousands of dollars' loss. From similar experiences to that one, I've adopted various tactics I use whenever I receive questioning from other teams. I focus on relaying the fact that these sometimes tedious but necessary validation requirements aren't coming from myself specifically but from a regulatory body, and we are all working toward the same goal at the site—to manufacture a safe product that is compliant with all relevant standards.

What do you wish you knew before choosing this career?

If I could go back and give my undergraduate self some pieces of advice that I have picked up throughout my career, I would start by emphasizing that problem-solving and critical-thinking skills are more highly regarded than technical knowledge for my specific position. I want to bring your attention to the fact that I included the phrase "for my specific position" because I am not in a research and development role, which depends on academic material much more significantly than my role as a validation engineer does. In this industry, almost everything can be found in some standard or guidance document released by regulatory organizations. The important aspect is understanding how to apply that documentation to whatever project you may be working on or issue you may be facing. Unfortunately, they don't teach you that in school, but they do allow you to practice building up your critical-thinking skills, which will help you address these aspects in the future.

While working in the industry, I quickly learned that being transparent is a better mantra to adopt than pretending to know something. There's no reason to be worried about how your lack of knowledge on one subject will make you look; it's important to work towards gaining that knowledge so you can store it away if the same topic comes up again in the future. It's not difficult to portray that you're willing to educate yourself and get back to that colleague on a specific question; it's all about the tone in which you say it. This applies across all different aspects of the corporate field as well, and not only in situations where you need to be transparent about your knowledge base. Always be clear with your intended remediation plan when dealing with these "stickier" situations.

Can you share additional ways you promote women in STEM?

In January of 2022, I started a podcast called *My Best Friend's an Engineer,* that I cohost with another engineer. This podcast focuses on discussing aspects of being a woman in the STEM field. We give advice from what we learned during engineering school, as well as our everyday experiences working in the industry. We also focus on inviting other women in STEM to speak on the podcast, which allows us to show different backgrounds, perspectives, and experiences to our listeners. In addition to the podcast, I post content about being a woman in engineering on my Instagram account, @EngineerLexi.

Connect with Lexi:

THE SPACE LATINA

Contributed by Zaida Hernandez:
Mechanical Engineer at NASA

Can you tell me about yourself?

I am a first-generation American in the U.S. My family is originally from El Salvador, C.A., and I am a Texas native.

Since childhood, I have always had a fascination with space. On family trips to El Salvador, I would try to count all of the stars in the night sky! When I was a senior in high school, I applied for a NASA internship, something I thought might be a bit out of reach, but my love for space wanted me to take a chance. Little did I know that it would change my life. I was accepted, and I fell in love with the space industry. My career goal was to work in the space industry as an engineer, and I can proudly say that I was able to reach that goal.

I feel very fortunate to work in the space industry. I work to guide others who are interested in STEM (Science, Technology, Engineering, and Math), and/or working in the aerospace industry. I love to participate in STEM outreach events and speak to students about the importance of careers in STEM. I recently published a bilingual children's coloring book called *Space Espacio,* because I believe biculturalism is important and representation in literacy matters. I want to help bridge a gap in STEM resources for bilingual children.

I graduated from the University of Houston with a Bachelor's of Science degree in Mechanical Engineering and a Master's of Industrial Engineering.

Why did you choose your field of engineering?

I am a mechanical engineer, and although I did consider other branches of engineering, mechanical engineering (in my opinion) gave me more options for careers. Through my high school internship at NASA, I got "real-world" exposure to engineering and had the opportunity to do some on-the-job training with aerospace, electrical, and mechanical engineers. Here I learned that my interest was in mechanical engineering, which ranges from structure analysis, to thermal design, to mechanism testing, and everything in between.

Can you describe the journey to get to your current role?

I work at NASA Johnson Space Center in their engineering division. I work on spacecraft to get humans back on the moon. I started my journey there as a high school student intern!

After that first internship, I continued to apply for internships and was accepted into the NASA Pathways Program, which is one of NASA's pipelines for full-time employment. After successfully completing this program, I was offered a full-time position.

What are the responsibilities in your current position?

I support the Orion Program at NASA. Orion is the name of the spacecraft on which the astronauts will travel to the moon under the Artemis Program. My group works on thermal systems, which includes analysis, design, testing, and management of the vehicle. I've had different roles before this one where I would conduct analysis and simulations of a module on the International Space Station, and help design the thermal system of a lunar rover.

Can you share some highlights of your career?

One of my biggest accomplishments is being recognized as a Spaceflight Awareness Trailblazer. This is an award given to early-career professionals in the aerospace industry who demonstrate creative and innovative solutions in human spaceflight.

Another highlight of my career was being recognized by various media outlets, including Hispanic media giants Telemundo and Univision.

What was more challenging than you had anticipated?

Although I was an honor student in high school, it was very difficult to transition to the university level. I was having a hard time balancing my internship, schoolwork, organization, and personal events. I struggled in a few classes. I even dropped a class, didn't do great in others, and was put on academic probation. I was at risk of losing my academic merit scholarships, which was very stressful. I went from feeling confident and accomplished in high school to feeling confused and lost in college. I had to think through how I would continue to pursue my goals despite the hard times.

The turning point came for me when I decided to get my life together, set priorities, focus on myself, and get through my obstacles. With a lot of dedication, some nights of little sleep, a lot of support from family and friends, and a lot of tutoring, I was able to get my grades back up and graduate with honors. That was my biggest victory. This was a slow process, but I had to remind myself that I was not going to allow anyone or anything to stand in the way of my goals and dreams.

What do you wish you knew before choosing this career?

I knew that by going into engineering I was going into a male-dominated career, but I didn't realize what that truly would be like until I was there. Most of my professors were men; the teacher assistants were men, and my peers were mostly men. I had to get used to always being the minority in the room. But as much as there was sometimes a "boys' club," I did find allies through student organizations. I was active in the Society of Hispanic Professional Engineers as well as the Society of Women Engineers, and those were both key in my engineering student life.

I wish I knew that everyone's journey is different, and that for many, engineering isn't a four-year degree. Some students change their major. Others, like me, participate in many internships that postpone our graduation date, while others are working and going to school.

That is all okay. I think the main thing to know is that engineering does require hard work, and as long as you are willing to put in the effort, you can make it to the end.

Can you share additional ways you promote women in STEM?

Through social media, I have been able to meet many wonderful women in STEM, some already in their careers and others still students. Through my page @thespacelatina, LinkedIn, and my website, I've been able to share my journey and also provide mentoring to women and young girls. I've spoken on podcasts, participated in STEM marathons, and volunteered in STEM events in hopes of helping at least one student. Perhaps I can provide them with a tidbit of information they may not know, or a different perspective that could help make their journey easier. At a minimum, I hope to provide the encouragement of having been in their shoes and getting through engineering, and they can do it too.

Is there anything else you would like to share?

My academic and career paths were not always clear. Because I was interested in graphics, I enrolled in engineering and architectural graphics courses at school. Originally, I considered architecture, but I also went to an all-girls engineering camp to learn about what engineering was and the different types of engineering. That led me to focus on engineering instead, which, looking back, was very important to my path. I realized that I liked making and creating! For this reason, I declared my major in engineering as I began to explore life after high school.

I graduated in the top 10 percent of my class and was very proud of that achievement, but it was not easy. I spent many hours preparing and figuring out the best approach. To help me study for the SAT and improve my score, I bought an SAT practice book. The goal was to practice and get used to the timed portions of the test. This helped me improve my score the second time around.

Connect with Zaida:

I WAS BORN TO BE AN ENGINEER

Contributed by Mandy Zhao:
Senior Compliance Engineer at TÜV SÜD America

Can you tell me about yourself?

Hi, my name is Mandy Zhao. I was born in China to an engineering family. Both my parents were mechanical engineers. Growing up, I listened to my parents discuss technical questions at the dinner table and watched them exchange opinions on drawings of the components designed for the system. It was quite natural for me to choose engineering as my future career field.

I have both my master's and bachelor's degrees in electrical engineering.

Why did you choose this field of engineering?

From a very young age, I showed strong interests in math and science, and during grade school, STEM (Science, Technology, Engineering, and Math)-related subjects were always my favorite subjects. So, I guess I was born to be an engineer, and I just followed my heart.

Where do you work and what was the journey like to get there?

After graduating from college in China, I entered a research institute as an assistant electronic engineer and worked there for six years. The focus was on the safety testing of telecommunication instruments and electrical toys. After that job, I was able to start my career as a safety compliance engineer at TÜV SÜD in Guangzhou, China. Being a member of TÜV SÜD, feels like having grown up in this big family.

I moved to the U.S. in 2003. While continuing with my career, I also pursued more advanced education, and received my master's degree in electrical engineering in 2007.

What are the responsibilities in your current role?

I have been working with TÜV SÜD for twenty-two years! Thanks to my company, I have had a lot of room and opportunities to grow. I started as a compliance engineer in product safety, and step by step, I became a senior engineer, technical certifier, technical lead, and senior product specialist. I also got the chance to explore different categories. I am now focusing more on the medical device safety aspect.

What was more challenging than you had anticipated?

I would not say that it is more challenging than I anticipated, but being in the STEM field does require a lot of effort. As an engineer, you have to constantly study and improve yourself to keep up with the rapid pace of technological developments. You need to keep an open mind to accept and absorb new things.

Can you share additional ways you promote women in STEM?

I mentor high school girls who are interested in the STEM field, and I am also an event speaker.

Is there anything else you would like to share with the reader?

Twenty years ago, sitting in my graduate school classroom, there were usually only a couple of girls in a thirty-student class. Now, I am more than happy to see more and more brilliant girls choose the STEM field as their career path. I would like to say to these girls: stereotypes shall not be the roadblock. Follow your heart, make sufficient effort, and you will thrive.

Connect with Mandy:

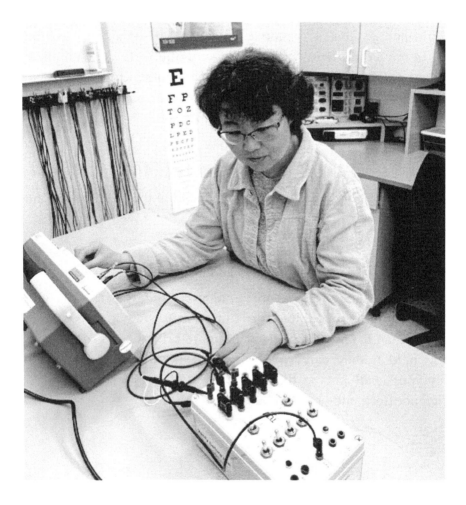

A LIFE IN PROBLEM SOLVING

Contributed by Dr. Susan Rogers:
Audio Engineer and Educator at Berklee College of Music

Can you tell me about yourself?

I grew up in Anaheim, California, during the 1950s and 1960s. My family was working-class, so there was never any serious talk about college. I was crazy about music, like a lot of kids. Often, when a parent has a child who's really interested in music, they sign them up for music lessons. My parents did that, but it wasn't a good fit for me. I took piano lessons, but I had no interest in it at all. What I loved was listening to the radio.

When I had a bit of money, I would buy records. My favorite fantasy was picturing the musicians playing live. I never imagined myself as a performer. Well, every kid does that a little, but it was never a fantasy I could sustain. My childhood was complicated because my mother got ill when I was eight years old, and she passed away when I was fourteen. Because I was the oldest, and the only girl in the family, I had adult responsibilities as a teenager. This included cooking, cleaning, ironing, all that kind of stuff. It didn't seem like my future was going to intersect with music in any way.

Eventually my dad got remarried, and my stepmother preferred to be the only woman in the house. So, at seventeen, I escaped with a boyfriend who was twenty-one years old. I thought *Okay, this is it, I am going to have an independent life.* So, I dropped out of high school and got married. That was a mistake, and that decision hurt me for years, because the guy I married became physically abusive. You hear stories of women who stay in relationships like that, and the reasons are complex. But I hated him, and I couldn't wait to get out.

Finally, when I was twenty-one years old, I escaped, again. I moved to Hollywood with a roommate. She and I set out on path's that were musically related, but both very different. Her big goal in life was to meet and marry a rockstar, and my big goal was to contribute to records—to somehow be a part of records and help bring the music I loved into the world.

Why did you choose your field of engineering?

When I started out in 1978, I pored over the credits on the back of album covers, looking for a woman's name. I saw Leslie Ann Jones and Peggy McCreary. If I hadn't seen those names, I would not have known that an engineering career was possible for a woman (it helps to know that you're not an oddball). I knew in my bones that I wasn't the sort of person who would be a record executive or a manager. I didn't want to work at a record label. Although I wasn't business-minded, I was technically adept. I realized, I could enter the music industry through a door where my gender made no difference: being an audio technician, repairing consoles and tape machines.

To make that happen, I bought volumes and volumes of books on basic electronics, and then I studied them all. Even though I was self-taught, with a lot of practice, soon enough, I got a job in Hollywood as an audio trainee with a company called Audio Industries Corporation. Later, I worked as the maintenance tech in a studio owned by Crosby, Stills & Nash.

How did you get to where you are?

In 1983, I got a phone call that changed my life. A former boyfriend said that Prince, my favorite artist in the world, was looking for an audio technician. I knew in that instant, that I would get that job! That was when my career began to accelerate. I joined Prince in Minnesota as his technician, working on *Purple Rain*—the album, movie, and tour! From there, Prince moved me into the engineering role, and I worked with him for five years. During that time, I also worked on albums by Sheila E., The Time, Madhouse, and Apollonia 6 (his protege bands).

As his engineer, I toured all over the world. I made movies and videos, contributing to whatever he did. During that time with Prince, and his people, I felt like I had done fifteen years' worth of work in five years.

After I left him in '88, I came back to Los Angeles, and for the next 12 years I worked as an engineer for some producers, and as a mixer for others. Eventually, I was asked to produce records and had commercial success with Barenaked Ladies in the late '90s. With the royalties I earned from the *Stunt* album, I left the music business to get my high school diploma. After that, I earned a bachelor's degree in psychology and neuroscience from the University of Minnesota. I then went north to McGill University in Montreal, where I got my doctoral degree in music perception and cognition. In 2008, I answered the call to be a full-time professor at Berklee College of Music in Boston. I taught record production, music cognition, and psychoacoustics.

What were your responsibilities?

Being an audio technician can be very difficult, but I really embraced it. You need a working knowledge of basic electronics. You need to follow the schematics and you need to be able to do component-level repairs. For example, if you pull a channel strip out of a recording console, you need to trace the signal to see where it gets lost and figure out which component has gone bad. The same is true with tape machines. Nowadays, in the modern era of programmable integrated circuits, you don't have to do as many component-level repairs as you once did, but a technician still needs a lot of training. It helps to go to college, but I was self-taught, and I did fine. Of course, you're going to have experts in the field who will train you as well. Technicians are not in the studio when musicians are playing and making a record, your job is to be there in the off-hours, making sure that everything works properly. From headphones and loudspeakers all the way to repairing cables, you tackle the biggest and smallest jobs in the studio.

When I worked as a technician, I was the only one capable of keeping the session running if something went down. That knowledge gives an engineer confidence. Typically, engineers start out as someone's apprentice (when an experienced engineer takes you under their wing). That can be harder for women because a lot of male engineers don't necessarily look at a young woman and think, S*he could have the same career that I have*. Starting as a technician meant I didn't have to go through the apprenticeship door and be an assistant engineer. The engineering role is more artistic than a technician's, because you make stylistic choices about sound sculpting and how your sound design serves a particular auditory scene. Typically, the producer chooses an engineer based on their artistry as well as their skill, and the kinds of sounds they are really good at getting.

Can you share some highlights of your career?

My most recent accomplishment is the book I've written on music listening, which was published in September 2022. It is called *This is What it Sounds Like*. My book is written from three perspectives: that of a record maker, from my Ph.D. work in music perception and cognition, and from the perspective of the general music lover, which is what I am. There's a chapter in there (chapter 8) that steps away from neuroscience and is devoted entirely to what music sounds like to a record producer. I chose the topic of music listening, because I'm not a musician, and I was invited to write a book, I wrote the book on music listening. It's doing well—getting a lot of reviews. I think that's a pretty good accomplishment.

What was more challenging than you had anticipated?

In hindsight, the most challenging thing outside of the day-to-day stuff was giving myself enough credit for what I did well and pushing myself to improve where necessary. It's hard to be a great service technician. Sometimes the problems are easy, but other problems are gut-wrenchingly difficult to solve. Same thing with being an engineer—sometimes it goes easily, other times you have a lousy day where you struggle to get the sound right. As a mixer or producer, you will have your good and bad days. Every single one of these roles is challenging. I wish I could have had better awareness and knowledge of my strengths and weaknesses. I only saw my RMS* level of performance skills as average. In many ways, I was stronger

---

* RMS is the abbreviation for Root, Mean, Square and is a math term that is the "a meaningful way of calculating the average of values over a period of time." From https://www.sweetwater.com/insync/rms-root-mean-square

and better than I gave myself credit for. That was probably the biggest challenge.

What do you wish you knew before choosing this career?

One general thing beginners often don't realize, is that it's going to take a *long* time to make a career. If you start in your early twenties, you can expect to have made a name for yourself in your thirties. Record-making skills take a long time to master because you don't always get the opportunity to practice them. Statistically speaking, only a small percentage of the work we do gets widespread attention, so we need to do a lot, in order to improve our chances of success. When you're in between projects, long stretches can go by without an opportunity to hone in on your skill set. I wish I had anticipated that. It was a long time before I really felt confident in my skills. The other thing to learn is that to achieve mastery, you'll have to make a lot of sacrifices. Music is typically performed at night. So, your nightlife and your social life won't be the same as those of someone who is working a nine-to-five job.

They tell me that record-making today doesn't require the same brutal hours that it used to when we worked with analog tapes. Engineers today may make fewer sacrifices, but the fact still remains, there will be sessions at night. You'll have to cancel a date, change your plans or not visit your family during a holiday. Things like that.

The third thing to consider, which women must consider more rigorously than men, is that if your career takes off in your late twenties or early thirties, you must think seriously about whether or not you want to have kids. Women can't reproduce into their seventies the way some men can. In our thirties, we may face tough choices about reproduction and raising kids.

Are there additional ways you promote women in STEM?

I accept every offer to talk about, write about, and teach women in the music industry! I think it's the older generation's duty to help the younger generation whenever we can.

Is there anything else you would like to share?

I heard a sentence that changed my life when I first started on this journey in the music business. I was working as a receptionist, in a tiny one-room schoolhouse in Hollywood, called the University of Sound Arts. One day, a student asked a teacher how to make sure he landed a job in this business. The teacher said, "If you always want a job, become a maintenance tech. You will always work." That sentence changed my life, and that sentence is still true today.

Engineering is considered the more glamorous role, but technicians are necessary and valuable. You don't have to worry about being rejected for your gender. You shouldn't always have to worry about gender, but in truth, when a crew is assembled to make a record, oftentimes it is (still) an all-male crew. When a woman is included, it changes the temperature in the room.

If I were to start my career today, one of my top priorities would be to set up a home studio. I'd spend every single day practicing my skill set, whether that was beat making, time alignment, or sample replacement. If you're going to call yourself a record maker, you need to be recording every day. It is also important to be a student of the game. As my mentor, the Producer Tony Berg, said, "Your knowledge needs to be encyclopedic in the sub-genre of music in which you do most of your work."

Imagine you're standing at the foot of a very tall mountain. You want to get to the peak of that mountain. But before you start climbing, you must consider the various routes. You can follow the well-trodden path, with the long line in front of it, or you can carve your own path and face more unknowns. Once you start climbing, you'll get to a higher vantage point, and then you can ask yourself, *Is this the right mountain for me?*

No matter what style it is, you need to know the bands that pioneered that path and who is doing innovative work today. This helps you predict which kinds of records are likely to be successful tomorrow (in that sub-genre). You need to be a student of the game.

In order to do that, it helps to know the important websites and music journalists—the people who are influencing the conversation. Know the field in which you wish to work, have role models in that field, and know the great record makers whose careers you'd like to emulate.

Dear Reader,

I hope you found these stories enlightening and motivating. As you can see from the compilation of stories gathered in this book, every STEM journey looks different. There are no limits to what you can achieve. Do not let what has happened to you, or what people have told you, define where you are going in the future. When you face difficulties, and you will, I want to encourage you to persevere in the face of adversity and see those challenges as opportunities to improve. It was how we chose to respond to the difficulties in life that made us stronger. When you step out into the uncomfortable, you will see a new side of yourself, and often there are so many opportunities on the other side of that decision waiting for you. Remember, you have everything within you to fulfill the dreams in your heart. With hard work and determination, anything is possible.

Made in the USA
Coppell, TX
06 May 2025

49065019R00115